GMOs Decoded

Food, Health, and the Environment
Series Editor: Robert Gottlieb, Henry R. Luce Professor of Urban and Environmental Policy, Occidental College

Keith Douglass Warner, *Agroecology in Action: Extending Alternative Agriculture through Social Networks*

Christopher M. Bacon, V. Ernesto Méndez, Stephen R. Gliessman, David Goodman, and Jonathan A. Fox, eds., *Confronting the Coffee Crisis: Fair Trade, Sustainable Livelihoods and Ecosystems in Mexico and Central America*

Thomas A. Lyson, G. W. Stevenson, and Rick Welsh, eds., *Food and the Mid-Level Farm: Renewing an Agriculture of the Middle*

Jennifer Clapp and Doris Fuchs, eds., *Corporate Power in Global Agrifood Governance*

Robert Gottlieb and Anupama Joshi, *Food Justice*

Jill Lindsey Harrison, *Pesticide Drift and the Pursuit of Environmental Justice*

Alison Alkon and Julian Agyeman, eds., *Cultivating Food Justice: Race, Class, and Sustainability*

Abby Kinchy, *Seeds, Science, and Struggle: The Global Politics of Transgenic Crops*

Sally K. Fairfax, Louise Nelson Dyble, Greig Tor Guthey, Lauren Gwin, Monica Moore, and Jennifer Sokolove, *California Cuisine and Just Food*

Brian K. Obach, *Organic Struggle: The Movement for Sustainable Agriculture in the U.S.*

Andrew Fisher, *Big Hunger: The Unholy Alliance between Corporate America and Anti-Hunger Groups*

Julian Agyeman, Caitlin Matthews, and Hannah Sobel, eds., *Food Trucks, Cultural Identity, and Social Justice: From Loncheras to Lobsta Love*

Sheldon Krimsky, *GMOs Decoded: A Skeptic's View of Genetically Modified Foods*

GMOs Decoded

A Skeptic's View of Genetically Modified Foods

Sheldon Krimsky

The MIT Press
Cambridge, Massachusetts
London, England

This book was set in Stone Serif by Westchester Publishing Services. Printed and bound in the United States of America.

Library of Congress Cataloging-in-Publication Data

Names: Krimsky, Sheldon, author.
Title: GMOs decoded : a skeptic's view of genetically modified foods / Sheldon Krimsky.
Description: Cambridge, MA : The MIT Press, [2019] | Series: Food, health, and the environment | Includes bibliographical references and index.
Identifiers: LCCN 2018016803 | ISBN 9780262039192 (hardcover : alk. paper)
Subjects: LCSH: Transgenic plants. | Crops--Genetic engineering. | Genetically modified foods.
Classification: LCC SB123.57 .K75 2019 | DDC 631.5/233--dc23
LC record available at https://lccn.loc.gov/2018016803

10   9   8   7   6   5   4   3   2   1

Contents

Series Foreword   vii
Foreword by Marion Nestle   ix
Acknowledgments   xiii
Introduction   xv

1   Traditional Plant Breeding   1
2   Molecular Breeding   9
3   Differences between Traditional and Molecular Breeding
     and Their Significance for Evaluating Crops   19
4   Early Products in Agricultural Biotechnology   29
5   Herbicide-Resistant Transgenic Crops   39
6   Disease-Resistant Transgenic Crops   49
7   Insect-Resistant Crops   57
8   Genetic Mechanisms and GMO Risk Assessment   67
9   Contested Viewpoints on the Health and Environmental
     Effects of GMOs   79
10   Labeling GMOs   93
11   The 2016 National Academies Study   103
12   The Promise and Protests of Golden Rice   119
13   Science Studies and the GMO Conflict   129
14   Conclusion   139

Notes   155
Index   183

Series Foreword

GMOs Decoded: A Skeptic's View of Genetically Modified Foods is the fourteenth book in the Food, Health, and the Environment series. The series explores the global and local dimensions of food systems and the issues of access, community well-being, and social, environmental, and food justice. Books in the series focus on how and where food is grown, manufactured, distributed, sold, and consumed. They address questions of power and control, social movements and organizing strategies, and the health, environmental, social, and economic factors embedded in food-system choices and outcomes. As this book demonstrates, the focus is not only on food security and well-being but also on economic, political, and cultural factors and regional, state, national, and international policy decisions. The Food, Health, and the Environment books therefore provide a window into the public debates, alternative and existing discourses, and multidisciplinary perspectives that have made food systems and their connections to health and the environment critically important subjects of study and for social and policy change.

Robert Gottlieb, Occidental College
Series Editor (gottlieb@oxy.edu).

Foreword
Marion Nestle

GMOs Decoded is a gift to anyone confused about genetically modified foods. In this latest addition to Sheldon Krimsky's prolific output of books about how societies interact with new technologies, he takes on a formidable challenge—to examine the science of GMOs as a basis for dealing with the ferocious politics they incite. I use the word "ferocious" advisedly. Positions about GMOs appear polarized to the point of outright hostility. Krimsky wants détente. If we understood the science better, we might be able to achieve more nuanced views of the risks and benefits of GMOs and of the genetic techniques used to create them.

To anyone familiar with Krimsky's previous and ongoing work, this book may come as a surprise. Trained in physics and philosophy, Krimsky is a sharp critic of the role of technology in society with particular interests in the ethical implications of genetics and biotechnology and in risk communication. I have long admired his work for its firm grounding in science and its clear delineation of the ways in which political, cultural, and other societal factors color perceptions of the safety and other risks of new technologies.

In *GMOs Decoded,* Krimsky takes a deep dive into the science of food biotechnology on its own, separate from issues related to how the science is used by the companies producing and profiting from GMOs, or is interpreted by proponents, critics, or the general public. An attempt to discuss the science of GMOs distinct from its politics may appear foolhardy, if not impossible, and Krimsky deserves much praise for taking this on.

I speak from experience. My book about food biotechnology, *Safe Food,* first published in 2003, began with a reference to C. P. Snow's two-culture problem—what Snow called the "gulf of incomprehension" between scientists and nonscientists over matters of technological risk. To greatly oversimplify: scientists argue that if GMOs are safe, they are fully accept-able and no further criticism is justified. But to nonscientists, safety is only one of many concerns about GMOs and not necessarily the most impor-tant. Holders of this broader view argue that even if GMOs are safe, they still may not be acceptable for reasons of ethics, social desirability, unfair distribution, nontransparent marketing, or inequitable and undemocratic control of the food supply.

What I observed in discussing those issues, and continue to observe, is the discounting of anything other than safety by extreme proponents of GMOs who perceive even the slightest question about nonsafety issues as an attack on the entire industry. This has forced critics of GMOs to focus on safety issues rather than the far less quantifiable issues of social desirability, pushing critics into positions that deny the possibility of any benefit of GMOs. The result: Snow's gulf of incomprehension.

Is the gulf bridgeable? Krimsky argues yes. From the perspective of the science, GMOs can either benefit or harm society. It behooves us all to try to understand what the science is about as a basis for coming to more informed opinions about the uses, value, and risks of GMOs—the politics.

But before getting to what Krimsky does in this book, I want to make one point about GMO politics: the GMO industry brought the polarization on itself. As I explained in *Safe Food*, the first GMO food, the FlavrSavr tomato, was intended to be marketed transparently as a triumph of American tech-nological achievement (I still have its label in my files). British supermar-kets sold tomato paste prominently labeled as genetically modified without opposition. That changed under industry pressure for nondisclosure. I was a member of the FDA's Food Advisory Committee in 1994 when the agency ruled against labeling GMOs, despite evidence that trust requires transpar-ency. The GMO industry fought labeling then, and won, and continues to spend fortunes fighting labeling.

The industry also promised that food biotechnology would feed the world and create new foods that would solve problems for the develop-ing world, such as those able to withstand poor soil conditions, excessive heat, and limited water. But instead, the industry concentrated on far more profitable insect- and herbicide-resistant first-world crops, a strategy criti-cized for the effects on society of its monoculture, patented seeds, heavy use of herbicides, herbicide-resistant weeds, and destruction of beneficial insects. The potential for foods with consumer benefits remains but has been largely unrealized. Trust requires fulfilled promises.

As readers of Krimsky's previous books surely know, he cares about such issues and others related to the politics of GMOs and their societal impact. But in this book, he wants readers to realize that the risks and benefits of GMOs depend on understanding the state of their science. Here, he takes on the scientific questions, one by one, clearly and dispassionately. This must have taken courage and a great deal of work. The science of GMOs is complicated and occurs at the level of molecules—DNA, RNA, and protein, of course, but also a host of less familiar molecules responsible for making genetic modifications work.

Fortunately, Krimsky writes clearly and succinctly about such things, his descriptions are easy to follow, and he defines terms as they are needed. He begins by asking whether GMOs differ from foods produced by traditional breeding and, if they do, whether the differences matter. He wants to know how GMOs affect health and the environment, whether they really are more productive than conventional crops, and whether they use fewer pesticides and herbicides. He asks if GMOs have nutritional or other benefits for consumers, and if and how they should be labeled. He deals with these questions in short chapters, along with others, that examine methods and risk assessment, review what expert committees say about such matters, and use Golden Rice as a case in point.

Krimsky's presentation of the divergent viewpoints about what the science means is exceptionally fair and even-handed. He insists that:

> This book is not about taking sides. My experience in studying scientific controversies that have public policy implications is that there are often truths, falsehoods, exaggerations, assumptions, fear-mongering, and uncertainties in the claims found on multiple sides of an issue. This book will succeed if it … demystifies the science and shows where there is consensus, honest disagreement, or unresolved uncertainty.

I think it succeeds admirably. Krimsky is straightforward about his own assessments. For example—spoiler alert—he concludes that evidence supports a qualitative difference between traditional and molecular breeding of food plants. On other questions, when he assesses the science as inconclusive, he says so. He wants readers to understand the complexity of the scientific issues, to be skeptical of arguments from either extreme in the debates, and to adopt nuanced positions on GMOs. Some aspects of GMOs may be worth opposing, but some may well be worth promoting. We all need to know the difference.

Krimsky tells us that in researching this book, his own positions became less polarized and more nuanced. Reading it, mine did too. Now it's your turn.

New York, August 2018

Acknowledgments

I am grateful for the helpful suggestions of Hyejin Lee, who read an early draft of the manuscript and with whom I collaborated on a study of Golden Rice; to Jonathan Latham for reviewing some chapters; and to Bill Freese for his input on sections of the book. I also wish to thank Tim Schwab for his collaboration on the National Academies of Sciences, Engineering, and Medicine's conflict-of-interest study. My discussions with Marion Nestle are always illuminating.

Introduction

Agricultural biotechnology's debut in commercial markets in the mid-1990s was followed by stops and starts. Some products were not successful in the consumer market or did not give farmers the edge they were predicted to have over traditional crops or their preexisting planting and harvesting techniques. Other products proved to be commercial successes, and some even became blockbusters in terms of market share. The term *GMO* stands for "genetically modified organism." It can apply to bacteria, viruses, plants, or animals. Other terms—such as *transgenic crops* or *genetically engineered (GE) crops*—are used specifically for plants. Some subsets of GMOs consist of GE crops. Other GMOs were developed for agriculture but are not crops, such as genetically engineered microorganisms. When the first GE crops were introduced, the seeds or plant parts (such as root cuttings for asexual reproduction) were patented and thus became protected intellectual property. The value of the first-generation GE crops to farmers was largely in producing greater yields and in reducing the dependency on chemical insecticides.

American science and technology has provided modern civilization with a vast cornucopia of consumer products and industrial innovations in manufacturing, energy production, transportation, and more recently digital technology. Rarely has a technological innovation sustained decades of public skepticism, opposition, and opprobrium across advanced economies. But this has been the fate of genetically modified crops and the processed foods derived from them. The United States agricultural sector has largely welcomed the innovations, but in the public mind, there remains oppositional thinking and skepticism. Even the terminology is in dispute.

At the very time that genetically modified crops and other GMOs were being developed for agriculture and the first field tests were being performed in preparation for commercialization, test plots in the United States, Europe, and Asia were destroyed by anti-GMO activists. Those opposed to GMOs were not interested in waiting for the test results. They believed that the entire enterprise of redesigning the germplasm of plants by the new techniques in molecular biology would result, at the very least, in poorer-quality food and, at most, in a dystopian agriculture leading to illness, nutritional deficiency, or environmental harm.

The idea of redesigning the biotic world was anticipated well before genes were discovered. Francis Bacon, the English polymath philosopher, scientist, and statesman, writing on the threshold of a new era of experimental science, constructed a fable about how the new sciences could create a world order of unimagined gifts. In his 1642 essay "The New Atlantis," Bacon's futuristic world consists of majestic architectural towers that are about a half mile in height (a precursor to Le Corbusier's "Radiant City" urban design), desalinization of sea water, and horticulture of unusual beauty and variation. He writes, "And we make by art … trees and flowers, to come up earlier or later than their seasons, and to come up and bear more speedily than by their natural course they do. We make them also by art greater much than their nature; and fruit greater and sweeter, and of different taste, smell, colour, and of figure, from their nature."[1] For Bacon, the existing biotic world represented the basic raw material or feedstock for transforming plants into more functional utilitarian objects. His vision was realized three hundred years later when gene splicing was discovered, and its applications in agriculture were set in motion. Although there are many examples throughout history of human modification of biological life, for the first time it could be done at the molecular level by moving genes across taxonomic systems.

Beginning in the mid-1970s, stakeholder groups established divisions well before there was credible risk assessment. Greenpeace launched an international campaign against field trials of genetically modified soybeans. Through books, blogs, white papers, and lobbying, other nongovernmental organizations met the prospect of genetically modified crops with different degrees of skepticism. The Union of Concerned Scientists set up a Washington, D.C., office run by two scientists who lobbied for better regulations against the unanticipated effects of releasing genetically modified crops into the environment. Groups like the Environmental Defense Fund questioned the U.S. decision to treat bioengineered crops similar to those produced by traditional breeding. The Foundation on Economic Trends opposed plant and human genetic modification as "playing God" with natural processes. Meanwhile, the National Academy of Sciences (currently the National Academies of Sciences, Engineering, and Medicine, NASEM) issued several reports stating that the use of recombinant DNA technology, whether for bacteria or plants, does not introduce any unique hazards.

In 1984, during the administration of President Ronald Reagan, an inter-agency working group called the Domestic Policy Council Working Group on Biotechnology, chaired by presidential science adviser George Keyworth, sought to provide "a sensible regulatory review process that will minimize the uncertainties and inefficiencies that can stifle innovation, and impair the competitiveness of U.S. industry."[2] The Biotechnology Working Group released its regulatory framework for biotechnology in December 1984. Its report provided the foundations for the oversight of biotechnology for decades to come. The key ideas were science-based regulations, the foster-ing of U.S. global competitive leadership, and internal and international harmonization of oversight, consistency, ease of regulatory burdens.

The U.S. oversight of biotechnology was placed in the hands of three agencies—the Food and Drug Administration (FDA), the U.S. Department of Agriculture (USDA), and the Environmental Protection Agency (EPA). No new laws were enacted. The agencies operated within their existing statutes. FDA issued its initial policy in 1992[3] and reaffirmed it in 2000,[4] asserting that transferred genetic material into crops are generally regarded as safe (GRAS): "FDA is not altering its view, as set forth in the 1992 policy, that there is unlikely to be a safety question sufficient to question the pre-sumed GRAS status of the proteins (typically enzymes) produced from the transferred genetic material or of substances produced by the action of the introduced enzymes."[5]

Despite the U.S. government's general approval of genetically modified crops and the extensive planting of a few staples like corn and soybeans, segments of the general public (including members of the scientific com-munity) continued to remain skeptical over the human health and environ-mental consequences of the new plant products. This skepticism was pitted against the strong affirmation of genetically engineered crops by scientific elites and major professional associations, such as the National Academies.

In industrialized societies, where science has acquired the position as the final authority over claims for generalized empirical knowledge, some segments of society are skeptical or even adversarial over what appears to be a strong scientific consensus among elites over GMOs. In the United States, that consensus covers the safety of genetically modified crops for human and animal health and their environmental impacts. Questions about product risk are largely, if not exclusively, viewed as lying within the purview of science. Although some scientists have declared that the debate over GMOs is over,[6] other scientists declare with equal confidence that there is no scientific consensus on GMO safety.[7]

Social scientists have studied public departures from entrenched scientific claims. A number of explanations can account for these departures. Among them are that not all scientific claims are exclusively fact-based and devoid of normative or value assumptions. When such assumptions or presuppositions form part of the edifice of a scientific explanation, it is neither unreasonable nor irrational for there to exist disagreements because the value judgments are not premised exclusively on scientific authority.

A second explanation that accounts for differences between public views and a scientific plurality is that nonscientists appeal to sources other than science for establishing or reinforcing their beliefs. Folk wisdom, religion, family traditions, alternative news outlets, and new age alternatives to allopathic medicine are among the sources of nonscientific beliefs. Social psychologists have used the term *confirmation bias* to describe the tendency to see new ideas or established scientific claims as confirming an existing belief.

Third, those individuals who are inclined to follow scientific advice exclusively on matters of risk and health benefits may accept the knowledge claims or statements highlighting uncertainty by outlier scientists who publish articles supporting views that fall outside the mainstream. The history of science teaches us that minority positions sometimes become validated and should not be discarded at the outset, especially when questions remain unresolved.

In this book, I accept as a starting position that in the United States scientists are largely supportive of the GMOs that currently are planted and consumed. Based on published statements from professional societies and the scientific literature, any concerns over the human health or environmental effects of this new generation of agricultural products have not been any greater than those of traditionally bred crops. For example, the National Academies of Sciences, Engineering, and Medicine's most recent and comprehensive 606-page report on genetically engineered crops finds that "the research that has been conducted in studies with animals and on chemical composition of GE food reveals no differences that would implicate a higher risk to human health from eating GE foods than from eating their non-GE counterparts."[8]

But the viewpoints over GMOs cannot be divided into a simple polarity. There is a spectrum of positions. Even some mainstream scientists are not inclined to embrace GMOs uncritically and without some caveats. Moreover, the research is still evolving. In order to give a voice to the wide spectrum of viewpoints, I explore the skepticism over GE crops held by different public groups, nongovernmental organizations, and scientists who do not always share the same views on risk and environmental impacts.

This book addresses the core issues of agricultural biotechnology through topics that have been widely discussed and debated but that remain unresolved in the minds of some and resolved by others. But the resolution of the issues has been contested by different stakeholder groups. This volume explores those differences to determine the extent to which the fault lines of disagreement exist among scientists or between the scientific community and the popular culture. My investigation seeks to understand why GMOs are banned or restricted in some countries and welcomed by others. In the inquiry, I examine the role of science in addressing the risks and benefits of GE crops. Without oversimplifying the science, the book is written to help nonexperts understand disagreements about the potential of GMOs to transform agriculture either favorably or unfavorably. Does it represent progress or peril?

For more than two decades, the GMO debates have centered on a series of recurring questions. The chapters are intended to address and respond to these issues:

1. How does traditional plant breeding compare with the plant breeding taking place in biotechnology, which I call *molecular breeding*? What are the differences and similarities? Is genetic engineering a continuation of the plant breeding that humans have practiced for thousands of years? Or is it qualitatively different and, if so, by what criteria?

2. What is known about the health assessment of genetically modified crops? How can we account for differences among scientific studies on animal feeding experiments? Are such studies appropriate for evaluating the health effects of genetically engineered crops? What, if anything, do those experiments reveal about whether GMOs are safe to eat directly

or be included in the food chain of processed food? If those experi-ments are not appropriate, how else are GE crops evaluated? Are more or greater risks (health or environmental) involved in transferring genes into plants from widely divergent species (such as across genus, families, and even kingdoms) than through intraspecies gene transfer?

3. What are the arguments that the regulatory oversight of bioengineered crops should or should not be stricter than the oversight for traditional crop breeding? What is the distinction between process-based and product-based regulation? Does it make sense to have special regula-tions for a process—namely, the use of gene splicing to create GMOs?

4. What evidence, if any, is there that genetically modified crops are more productive (produce greater yields) than traditionally bred crops? This question bears on whether GMOs will contribute to a greater supply of food given the same quantity of seeds and other production inputs (such as fertilizer, water, and land).

5. What distinctions are there, if any, between the environmental impacts of GMOs and traditionally bred crops? Are some GMOs more favorable environmentally than traditional crops? Do GMOs have any unique impacts on biodiversity, beyond the impacts of traditional crops? Will GMOs contribute or become an obstacle to sustainable agriculture?

6. Have any commercialized GMO crops been designed to improve a crop's nutritional quality, flavor, or other attributes valued by consumers or public health advocates (such as through biofortification)?

7. What are the critical issues regarding the demands for and against man-datory labeling of GMOs? Is there a rational basis for labeling? How does the European Union compare with the United States in regard to label-ing genetically engineered crops and GMO food? What federal or state initiatives have been taken to label GMO crops in the consumer market?

Skepticism and ardent support for GMOs can be found across many types of organizations. These groups typically cite the scientific studies or public surveys that support their opinions. This book is not about taking sides. My experience in studying scientific controversies that have public policy implications is that there are often truths, falsehoods, exaggerations, assumptions, fear-mongering, and uncertainties in the claims found on multiple sides of an issue. This book will succeed if it lays out the claims and counterclaims and points to supporting arguments in a manner that demys-tifies the science and shows where there is consensus, honest disagreement, or unresolved uncertainty. The reader who gains a deeper understanding of the nature of the debate and the fault lines that divide communities will be in a better position to make an informed judgment about the previous questions.

The published literature on GMOs covers thousands of works, including journal articles, books, government reports, and NGO studies. Any author has to make a selection and connect dots among the vast reservoir of knowledge claims. My approach is to depend foremost on refereed journal articles, academic books that have been vetted by other scholars before publication, and reports from highly recognized journals, government agencies, and professional societies. I do not cherry-pick the science that supports a predetermined position. Science is a meritocracy but one where there is no forced hierarchy of opinion. Where there is honest controversy within the meritocracy, it can be found in the published scientific literature. Lack of published controversy within the canonical scientific literature can be a sign of consensus over a specific question.

We also have to be aware that scientists, even while publishing in the best journals, can carry hidden biases that they hold consciously or unconsciously. These biases can be reflected in financial conflicts of interest that may or may not be disclosed in their published papers. When scientists hold an equity interest in companies that are poised to benefit from a discovery, there is no longer an objective playing field between benefits and risks. The expectation of personal reward can diminish concerns about untoward consequences and can enhance positive interpretations of results. Unconscious biases can drive a weighted interpretation (or misinterpretation) of observational data from well-designed research or a preferential selection of studies on which to ground a preferred conclusion.

After this introduction, I begin by discussing traditional and molecular breeding. Chapter 1, on traditional plant breeding, covers cross-breeding, hybridization, and mutagenesis. Chapter 2 examines the methods for developing bioengineered crops, which I call *molecular breeding*. Here I explain the composition of "foreign" gene constructs that are transferred into the plant genome and discuss both anticipated and unanticipated outcomes. Chapter 3 discusses the distinctive features of molecular and traditional breeding and the ways that those distinctions in plant breeding affect questions of risks and benefits.

Chapters 4 to 7 examine the early and later commercial products of molecular breeding, including the Flavor Savr tomato and disease-, herbicide-, and insect-resistant crops. The reader is introduced to genetic mechanisms for developing genetically engineered crops and their relevance to risk assessment.

Chapter 8 delves into genetic mechanisms for molecular breeding and the ways that they enter into risk assessment. It contrasts the Lego model with the ecosystem model of the plant genome and the ways that they inform risk assessment of GMOs. In chapter 9, contested scientific viewpoints on the health effects of genetically engineered crops are explored. Whether GE crops should be labeled is examined in chapter 10. A critically important study by the National Academies of Sciences, Engineering, and Medicine is discussed and analyzed in chapter 11. Chapter 12 looks at the history, significance, and current status of the first biofortified GMO crop, called Golden Rice.

For chapter 13, I review the literature covering the social studies of science, also called science and technology studies (STS), to investigate how STS scholars understand stakeholder conflicts over GMOs and whether these conflicts can be resolved by a consensus among scientists. My conclusion on the current state of molecular breeding and the social angst over GMOs is the subject of chapter 14, where I return to answer the questions raised in the introduction.

The approach I have taken for exploring the vast body of scientific and policy literature begins by drawing from a representative sample (hundreds of studies and reports) of the tens of thousands of biological and social science publications on genetically engineered crops. A Boolean search of three science databases using the key words [(GMOs) OR (genetically modified organisms) OR (bioengineered crops) OR (genetically engineered crops) OR (genetically modified crops)] yielded 4,920, 5,130, and 49,658 publications in *Web of Science, Biological and Agricultural Index*, and *Pub Med* respectively.[9] Additional searches were carried out between 2016 and 2017 to capture current publications. My literature search was narrowed by topic areas, and special attention was given to contested findings and viewpoints in areas represented in the book's chapters, allowing the reader to understand the presuppositions and evidence behind the conclusions drawn. The approach I have chosen will succeed if it allows the readers to understand why there remains disagreement about the health, environmental, political, and social impacts of GMOs.

1 Traditional Plant Breeding

From archeological findings, we have learned that agriculture began in the Fertile Crescent of the Middle East somewhere around ten thousand years ago.[1] Named for its quarter moon shape, the Fertile Crescent is the region in the Middle East referred to as *Mesopotamia* (Greek for "between two rivers"), and it flanks the Tigris and Euphrates Rivers, covering parts of modern-day southern Iraq, Syria, Lebanon, Jordan, Israel, and northern Egypt. There is evidence that barley, wheat, and lentils were domesticated during that period. Human selection of desired crops for reproduction, as noted by George Acquaah, is "the act of discriminating among biological variation" in a plant population to select desirable cultivars.[2]

The earliest agricultural practices included tilling the soil, planting seeds, and harvesting crops. Human societies evolved from hunter gatherers to planters of seeds, which fostered stationary settlements around cultivated soils. Reflecting on the origins of conventional plant breeding, John Bingham wrote, "The first stage of plant breeding was selection of genetically varied crops before people knew what genetics was. There were variations in both cultivated and wild crops from which farmers could choose for desired traits. Since each farm ecological system was uniquely suitable for certain crop genetics, the farmer's role was to discover the best variant for his unique environment."[3]

Although it is generally understood from archeological discoveries that humans have fashioned tools for more than two million years, the development of human observation of and intelligence about how plant seeds can produce crops is fairly recent.[4] The earliest method of plant breeding was selection, defined as "the act of discriminating among biological variation to select desirable variants."[5] The traits sought were not unlike those that are sought today, such as resistance to disease, fruit size, color, and taste.

Certain variants that once were highly selected have been protected. The term *landrace*[6] refers to a cultivated variety with a distinctive identity that has evolved over centuries of farmer selection without formal crop improvement.[7]

The National Research Council (NRC), which is a part of the National Academies of Sciences, Engineering, and Medicine (NASEM), has classified two techniques for improving crops—selection and breeding. It does not view natural selection, which determines the survival of species, as a method of breeding. Incidental selection was eventually augmented by artificial selection. According to the NRC, "The earliest farmers selected plants having advantageous traits, such as those that bore the largest fruit or were the easiest to harvest. Perhaps through some rudimentary awareness that traits were passed from one generation to the next, the choicest plants and seeds were used to establish the next year's crop."[8] *Artificial selection*—the method used to narrow and control the available gene pool—provided an additional tool over incidental selection. In the former case, "a genetically heterogeneous population of plants is inspected and 'superior' individuals—plants with most desired traits, such as improved palatability and yield—are selected for continued propagation."[9] Thus, by selecting and isolating choice plants for cultivation, the early farmers were in essence influencing which plants would cross-pollinate. Through selection and isolation, they were narrowing, yet controlling, the available gene pool for each crop.

The next stage in plant breeding occurred when agricultural societies learned how to engage in plant reproduction. This stage in plant breeding was believed to have been practiced by the Assyrians and Babylonians around 700 BC, when farmers learned how to pollinate plants artificially by introducing pollen (male gametes) into the stigma (female reproductive part of a flower), circumventing natural (and thus random) pollination by wind or insects. What has been termed *conscious classical plant breeding* or *crossing* has resulted in many artificially pollinated crops, such as the palm and tomato.[10]

In 1984, the National Research Council defined *breeding* as hybridization where "farmers selected two plants and then crossed them to produce offspring having the desired traits of both parents. This was a trial and error process, however, since early plant breeders did not understand the genetic transmission of traits and could not predict the likely outcome of a particular cross."[11] This technique improved on the accidental crossing of sexually compatible crops because it involved human intervention in the sexual life of plants.

The crossing of different varieties or species to produce new ones, referred to as *deliberate hybridization*, arose as an outgrowth of the knowledge of plant reproduction. Rudolf Jakob Camerer, professor of natural philosophy at the University of Tübingen, Germany, reached a conclusion in 1694 that plants were made up of male and female parts and that pollen could fertilize the plant.[12]

Experimental science had begun to take hold in the 1600s. For plant breeding, a major advance was the introduction of botanical gardens in the 1800s. In these gardens, plants were carefully classified and observed by botanists, who could determine by experimental trials the precise plant characteristics they desired. No longer did they have to depend on the idiosyncrasies of individual farmer choices or natural fertilization. The botanic gardens represented the first step in scientific plant breeding and were the source of many new European crops.[13]

Working within the same species, plant breeders developed inbreeding and outbreeding techniques to create desired varieties. *Inbreeding* occurs when the male and female gametes of the same strain are bred back (*backcrossed*). This creates plants of stable homogeneous properties called *pure lines* (which may take more than one generation), where the offspring resemble the parental lines.

In *outbreeding*, male gametes (sperm) that are carried through pollen tubes fertilize female *gametes* (eggs inside the ovary) from different genetic backgrounds (*outcrossing*). For example, a domesticated plant can be outcrossed with a wild type, combining two genetically different individual plants from the same species. Outcrossing maximizes genetic variations in the variety, allowing the breeder a broader genetic stock from which to work.

A breeder may combine two different traits from two individual varieties to create a single plant, which possesses the two desired traits from each variety. Although this can happen randomly in nature, the more assured way to accomplish this is first to create two pure lines through inbreeding. Then the breeder cross-pollinates the two pure lines to obtain a hybrid with the two traits in one variety. This is called *hybrid seed technology*. For this to work, the pure lines must be reproductively compatible. Interspecies crossing between closely related species can take place naturally (by cross-pollination) or through human intervention (by induced pollination).

The next stage in plant breeding was advanced at the turn of the twentieth century when the field of genetics gained traction in plant biology. Two developments that advanced traditional breeding are the inducing of mutations in plants and the advancement of cell culture methods. Both introduced greater variability in the plant germ plasm for crop development. *Mutagenesis* in plants was induced on the gametes by radiation or chemical mutagens. The induction of mutations did not introduce foreign genes into the plant but rather modified the available germ plasm of the plant variety. Although the method could increase the variability of traits, it was a trial and error process that took years to reach a useful outcome. Plant reproductive cells were exposed to gamma rays, protons, neutrons, alpha particles. or beta particles to create mutations in the cells' genome. Scientists could then observe whether and how the mutations changed the plant's physical characteristics (*phenotype*). There was no way to predict the outcome other than by observing the results of countless mutations.

A second breeding process involved *culturing* plant cells. This was particularly useful when plants were not by nature reproductively compatible. Scientists learned how to overcome reproductive incompatibility by developing methods to induce reluctant stigmas to be fertilized by foreign pollen. Methods for circumventing cross-incompatibility included electrical currents, wire brushes, and heat treatment. After the embryo from the mechanical tools of hybridization was created, it required special treatment in culture. This process (called *embryo rescue*) involves extracting the immature hybrid embryos, which ordinarily would not survive, by growing them in a culture with nutrient solutions.[14]

After plant cells could be grown in culture, breeders observed spontaneous mutations in the cells, also called *somaclonal variations*. Scientists observed that the mutations provided potentially valuable variants for new varieties. The process is imprecise and consumes a lot of time.

Thus, traditional breeding includes many unnatural methods of plant reproduction. For example, *hybridization*, a process of reproductively uniting two different plant varieties, can be accomplished by crossing natural breeding barriers. One such method is *chromosome engineering*, where portions of plant chromosomes are combined through a process of chromosome translocation. However, this process does not easily remove detrimental genes that are carried along in the transferred chromosome.

Questions that have arisen in discussions among biologists include the following: Are there limits in crossing species boundaries beyond which conventional breeders cannot go to create new cultivars? Does biotechnology increase the range of hybridization over traditional breeding? If so, how far can it be extended? What is the significance of that extension?

Before I explore these questions, I pause to discuss how biologists classify organisms and what that classification reveals about the differences between traditional breeding and molecular breeding. The biological world is classified and ranked into nine categories. At the top of the ranking is life, under which fall all living organisms from viruses to primates. The classification system continues with domain, kingdom, phylum, class, order, family, genus, and species, where specificity increases from domain to species.

Thus, humans are a species and part of the family Hominidae, which also includes chimps, gorillas, and orangutans. Cabbage is a species that is part of the family Brassicaceae, which also includes horseradish, mustard, rutabaga, broccoli, Brussels sprouts, cabbage, and cauliflower. Mushroom is a species that is part of the family Morchellaceae, which includes 146 different species and ten genera. Finally, the species common pond amoeba is in the family Amoebidae, which includes the genera *Chaos*, *Entamoeba*, *Pelomyxa*, and *Amoeba*. The classifications genus and species are the only classes within the taxanomic classification where crosses in conventional breeding take place. Conventional breeding permits the movement of genetic material between different species or closely related genera.

Wide crosses are defined as "crosses between species of the same or different genera."[15] The greater the phylogenetic distance between the species, the more difficult it becomes to engage in a wide cross. These crosses, when they occur, are usually accomplished with artificial techniques. For example, plant breeders can get the pollen of species A to fertilize the egg of species B. Because the embryo may not be able to survive naturally, the breeders use the technique of embryo rescue, as previously noted, by removing the embryo from the hybrid seed and culturing it with plant hormones and nutrients. According to a National Research Council report, "Such embryo rescue is not considered genetic engineering, and it is not commonly used to derive new varieties directly, but is used instead as an intermediary step in transferring genes from distant, sexually incompatible relatives through

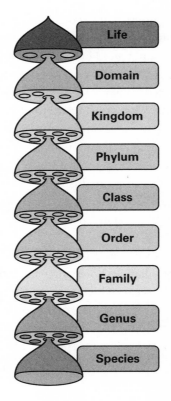

Figure 1.1
A taxonomy of living organisms

intermediate, partially compatible relatives of both the donor and recipient species."[16]

H. C. Sharma reports that "hybrids from crosses spanning wide taxonomic boundaries" have occurred when wheat pollinated with maize and produced embryos and when hybrids were created from wheat and sorghum or from rice and wheat. In these cases, he is referring to crosses between members of different subfamilies.[17] He notes that "crosses may be as wide as one can make them."[18] Wide crosses can, under unusual circumstances, cross two families. Most frequently, wide crosses can be induced among different species and less frequently across a genus. There is no evidence that wide crosses in natural processes or by the traditional non-natural methods of plant breeding can take place across groups in order, class, phylum, kingdom, and domain.

In his book *Hybrid: The History and Science of Plant Breeding*, Noel Kingsbury writes that

> Much of twentieth-century plant breeding has been concerned with getting pro-
> gressively less and less well-related relatives to cross, moving up to crosses between
> apparently close genera. Methods of overcoming cross-incompatibility by forcing
> reluctant stigmas to accept alien pollen have become increasingly bizarre (or des-
> perate): passing electrical currents across stigmas during pollination to get Brus-
> sels sprouts and savoy cabbage to cross, "mutilating"…cabbage stigmas with a
> wire brush, heat treating of lily…styles prior to pollinations.[19]

Artificial hybridization between plants of different species and plants
and organisms within different genera has expanded in the twentieth century. The methods of traditional breeding for creating crop varieties by
hybridization seem to have reached their limits. As Sharma notes: "the difficulty in obtaining wide hybrids increases with the phylogenetic difference
between the parental fauna involved."[20] Breeders also have used vegetative
propagation of stem cuttings. This has allowed them to produce clones of
plants, which do not create new traits or widen genetic variability. They
also have created new varieties by propagating plants in tissue culture with
the use of *protoplasts*—plant cells with the cell wall removed enzymatically.
The protoplasts can be combined or modified to regenerate a whole new
plant variety by transferring traits between species. This is called *somatic
hybridization* or *cell fusion*. The protoplasts derived from different plant cells
are fused together by electrical techniques. Also, by processes of electroporation or microinjection, DNA can be transferred to plant cells. *Electroporation* sends electrical impulses through the protoplast, which increases
the protoplast's membrane permeability and allows foreign DNA to enter.
Through *microinjection*, foreign DNA can be injected into the protoplasts.

Another method of modifying the genetics of crops is by changing the
number of chromosomes in a plant variety. This can be done by crossing a
plant with two sets of chromosomes (diploid) with a plant that has four sets
(tetraploid), resulting in a plant with three sets (triploid), such as a banana.[21]
On the down side, the triploid banana is sterile. Alternatively, one can use
chemicals such as colchicine to double the number of chromosomes.

Mutational breeding can use chemicals and radiation such as x-rays.
Plant cells or seeds are exposed to mutagens at doses that are nonlethal
but high enough to induce mutations resulting in new traits. The Food
and Agricultural Organization of the United Nations collaborates with the

International Atomic Energy Agency to maintain a mutant variety database online. From 1930 to 2014, it is estimated that more than 3,200 mutagenic plant varieties from 214 crops in 60 countries, including 1,000 varieties of staple crops, have been made available. Although mutational breeding changes the DNA of a plant cell, it is not classified as genetic engineering by regulatory agencies in the United States or Europe.

In conclusion, traditional or conventional breeding encompasses natural hybridization and selection, artificial hybridization among sexually compatible species within the same or different genera, the exposure of gametes to mutations with chemicals or radiation, the use of protoplast fusion that combines genes from the cells of two species, and chemically induced polyploidy (more than two paired sets of chromosomes).

Traditional breeding can be distinguished from biotechnology by the fact that the former operates at the cellular scale, whereas the latter operates at the molecular scale. As such, biotechnology has been aptly called "molecular plant breeding."[22] By operating at a molecular unit of analysis, scientists have been able to overcome many of the limits of genetic crosses previously discussed: "Traditional breeding focuses on individuals and populations and relies primarily on sexual reproduction to manipulate useful variability; in contrast, biotechnology focuses on the cellular and subcellular levels, capitalizing on the techniques of molecular biology both to generate and manipulate useful variability."[23] However, the line between traditional breeding and biotechnology has been blurred with technical advances. For instance, molecular-level techniques are applied to traditional breeding through *marker-assisted selection* or *quantitative trait loci*. Without inserting new genes, gene sequencing gives breeders an opportunity to screen out for desired crops more quickly than by trial and error, which then can be bred by traditional breeding methods. Also, gene editing by *CRISPR* (clustered regularly interspaced short palindromic repeats) can be used to make specific mutations in plants, reducing the uncertainty of mutations by chemicals and radiation. The next chapter explores the methods used in biotechnology to create new plant varieties by molecular breeding, comparing and contrasting those with traditional breeding techniques.

2 Molecular Breeding

New discoveries in gene splicing made during the 1970s evolved into research applications in plant genetic engineering in the 1980s and eventually to the commercialization of genetically engineered (GE) crops in the 1990s. With the development of libraries of restriction enzymes and DNA ligases, scientists had the basic tools for cutting and splicing DNA sequences from any biological source. The primary methods for *molecular breeding*[1]—that is, the creation of plant varieties by introducing into plant cells DNA from foreign sources by laboratory procedures involving recombinant DNA molecule technology—were now in place.

A six-stage process for molecular breeding was widely adopted:

Stage 1: The first stage involves identifying and isolating the foreign gene for transport into a crop, and it can be a long process. The purpose of transplanting a gene from another species into the plant is to endow the plant with a useful phenotype for agriculture, which can include greater efficiency in production, processing, or storage; higher nutritional content; and resistance to environmental threats (such as salt, weeds, and insects). With the development of recombinant DNA techniques, the gene can come from any organism in any species from any biological kingdom. The most popular genes transplanted into food crops have come from bacteria, viruses, or other plants.

Stage 2: In the second stage, the foreign gene is attached to other DNA segments to prepare a transgene complex with components that provide specific instructions to the host cell.

Stage 3: The third stage consists of preparing the host plant cell or tissue to receive the transgene complex in such a way that it can regenerate into whole plants. Generally, this involves removing a tissue piece from a plant, using it to prepare cell cultures, and maintaining those cultures under appropriate hormonal and nutrient conditions.

Stage 4: The fourth stage consists of selecting a delivery system for the transgene complex and its method of transfer into the host cell.

Stage 5: The fifth stage involves selecting the transgenic cells—that is, those cells that have taken up the foreign DNA from other non-transgenic cells within the cell culture and their regeneration into whole plants.

Stage 6: In the sixth and final stage, just prior to the introduction of the GMO into the marketplace, the transgenic plants are evaluated for the desired phenotype with a heritable genetic change and are examined for other characteristics, such as nontoxicity, nonallergenicity, and nutritional equivalence to the parental crop.

In stage 1, the molecular breeder identifies and selects a foreign gene for transfer into the host plant cells. Scientists choose a DNA sequence that encodes a protein from a biological source and that they want to see expressed in the crop plant. It could be a bacterial gene that would afford the plant with a trait like insect or herbicide resistance or a viral gene that would make the plant disease resistant. Other genes investigated include those that would endow the plant with abiotic traits like salt or stress tolerance. The genetically engineered crops that are designed to improve agricultural production by adding traits like insect resistance are termed "first-generation GMOs." In the first few decades of commercial applications of molecular breeding, scientists focused almost entirely on altering traits that were deemed useful for agricultural production.

After the gene is selected from another organism and the required DNA sequence is isolated (by restriction enzymes), it can be produced in large quantities by *polymerase chain reaction* (PCR). I use the term *transgene* to mean the foreign gene that is isolated for transfer to a new biological system.

Stage 2 is the construction of the transgene cassette. After the gene is isolated, it must be attached to several other DNA sequences that make up what is called a transgene cassette. In other words, the transgene is not transferred by itself. Other sequences are added that play important roles in getting the gene to function in its new environment. One of the DNA components added to the transgene is the *promoter sequence,* which is attached

Promoter	Transgene Gene from foreign source	Marker gene	Termination sequence

Figure 2.1

After a gene has been isolated and cloned (amplified in a bacterial vector), it must undergo several modifications before it can be effectively inserted into a plant. This simplified representation of a constructed transgene cassette contains the necessary components for successful integration and expression. The marker gene also has its own promoter and terminator.

Source: http://cls.casa.colostate.edu/transgeniccrops/how.html.

to the desired gene (figure 2.1). It provides the switching (on and off) mechanism to activate and regulate gene expression in the gene's new environment. A commonly used promoter, known as CAMV35S, comes from the cauliflower mosaic virus. It affords a high degree of expression to the foreign genes transplanted into the plant cells and enables gene activity throughout the life cycle of the plant in most tissues. Thus, a gene for an insecticidal protein will, in all likelihood, be expressed at all stages in the life cycle of the plant.

A second component in the transgene cassette is the *termination sequence* (also called a *stop codon*), which signals to the cellular machinery that the end of the DNA sequence has been reached. The termination codon is a nucleotide triplet that signals a termination of translation into proteins.

A third component in the transgene cassette is the *marker gene*. Its purpose is to provide the molecular breeder with a mechanism to select the plant cells that have successfully incorporated the transgene cassette. This represents stage 5 in the six-stage process. The marker is a DNA sequence that protects the transgenic plant cells from being destroyed by some external agent that is toxic to the cells without the marker gene. Commonly used markers include genes for herbicide resistance or antibiotic resistance. In the former case, after the transgene cassette has been transmitted to the cells, the cell culture is exposed to an herbicide. Only those cells with the herbicide-resistant marker will be viable. Where antibiotic resistance markers are used, the cells targeted for transformation are exposed to antibiotics, destroying the cells that do not possess the selection marker gene and leaving only the transgenic cells.

In stage 3, plant cells or tissues selected for receiving the transgene cassette are incubated in a medium known as *tissue culture*. The challenge for plant geneticists is to find the delivery systems to transport foreign genes into the cells or embryos of plants and then to ensure that the transported genes function effectively in their new host plant environment.

For stage 4, two classes of DNA delivery systems, sometimes referred to as *indirect* and *direct*, are generally used. Biology-based gene delivery systems, also called *vectors*, are bacterial plasmids and viruses that naturally carry DNA into plant cells. Although plants have evolved resistance to many types of infectious agents,[2] some viral or bacterial agents are invasive to certain plants. They enter the cells of the plant, use its biological machinery to replicate, and sometimes inflict damage to their host. When these vectors carry the transgenes, they are referred to as *indirect delivery systems*.

There are also nonbiological systems that use chemical and physical processes for transporting foreign DNA directly into plants (*direct delivery systems*). These systems do not depend on whether the organisms that provide the source of DNA are reproductively compatible (can naturally or even artificially reproduce or hybridize with the recipient organism). The ability of biotechnology to overcome sexual incompatibility, organismal barriers to hybridization, or grafting barriers between two species is what can be termed *molecular breeding*.

The most widely used biology-based genetic transformation system for plants is *Agrobacterium tumefaciens* (*A. tumefaciens*), a soil-dwelling bacterium that infects certain plants and causes crown gall disease. *A. tumefaciens*, like all bacteria, contains a chromosome that includes most of its genes and additionally a unique plasmid (the Ti plasmid), which is a circular ring of DNA with a smaller number of genes. The Ti plasmid, named for its tumor-inducing properties, contains vir genes and T-DNA (for transferred DNA). The function of the vir genes is to code for proteins that open a channel that allows the T-DNA segment of the Ti plasmid to pass into and infect plant cells. When *A. tumefaciens* infects plant cells, it deposits the T-DNA segment of its plasmid into the chromosome of the host plant. These take over the machinery of the cell to synthesize sugars that feed the bacteria.

A. tumefaciens cannot enter a healthy plant. But if a plant stem or root is mechanically damaged or biologically wounded, it emits a chemical signal that is picked up by the vir genes of the bacterium. The vir genes activate a series of events resulting in the transfer of the T-DNA sequence from the

Ti plasmid to the plant's chromosome. This remarkable signaling system is exquisitely tuned to the highly evolved bacterial symbiont. The T-DNA sequence of *A. tumefaciens* initiates a tumor in the plant called *crown gall tumorigenesis*.[3]

The discovery of how *A. tumefaciens* infects plants and transports the T-DNA segment causing crown gall disease was made in 1977 by Mary Chilton and colleagues. After the mechanism of disease transfer was understood and scientists saw that *A. tumefaciens* can transport its DNA segment into plants, they wondered whether replacing the genes on the bacterial plasmid with ones of their own choosing would allow the bacterium to transport foreign DNA into plant cells. When scientists replaced the genes, they removed the tumor-inducing gene from the Ti plasmid so that it would not cause disease in the plant.

To transport genes (the coding DNA) via the *A. tumefaciens* bacterial vector, scientists have to go through several steps. First, the target gene for transplantation (the transgene) has to be isolated. Second, a construct has to be developed consisting of four components—a promoter, the coding DNA or transgene itself, a terminator sequence, and a marker gene (as previously described).

The promoter DNA is necessary to activate the correct level of expression of the transgene in the cell. It is able to turn the gene on or off. As mentioned earlier, the most widely used promoter is a DNA sequence from the cauliflower mosaic virus called CaMV35S. It induces a high level of expression for the transgene and is active in almost all circumstances in the cell without stimuli,[4] but there are exceptions where the promoter does not work as advertised.[5] Finally, the termination sequence signals to the cellular machinery that the end of the gene has been reached.

The most common nonbiological mechanism used to transport foreign genes into plant cells is a projectile system referred to as a *gene gun*. Tiny gold or tungsten particles that have been coated with transgenic DNA cassette are propelled through the walls of the plant cells into the plant's chromosomes. The gene gun process is also known as *microprojectile bombardment* or *biolistics*. Typically, the projectiles are directed at cell cultures or immature embryos. Particle bombardment was first discussed for transgenic plants in 1987.[6]

Other direct methods used in transplanting foreign genes include *electroporation*, which involves the application of a strong electric field across cells and tissue. The electric field makes the cell membrane porous so that purified DNA can be taken up by the cell. Various other methods for opening up the cell wall have been reported in the literature. In silicon carbide mediated transformation (SCMT), silicon carbide fibers puncture the cell walls, allowing DNA to enter. The treatment of plant protoplasts with polyethylene glycol (PEG) also allows foreign DNA to enter the plant cells. And pollen tube pathway (PTP) is another important delivery system for transferring naked DNA into plant ovaries. Although several systems for integrating foreign DNA into the germ plasm of plants have been tried, *Agrobacterium tumefaciens* and biolistics are the most efficient and widely used systems.[7]

Molecular breeding, as described here, does not occur in nature. The transgene complexes with promoters, markers, and termination sequences are produced in a laboratory and transferred into plant cells by biological or mechanical vectors. As sophisticated as these breeding methods appear to be, they rarely produce the desired trait in normal plants. It takes many tries. As noted by Jonathan R. Latham, Allison K. Wilson, and Ricarda A. Steinbrecher, "even after selection, there are many reports of apparently normal transgenic plants exhibiting aberrant behavioural or biochemical characteristics ... [ranging from] altered nutrient or toxin levels to lower yields."[8] It is likely that transformation—that is, induced mutations or damage to host plant DNA caused by DNA introduction—are behind many of the unanticipated effects of molecular breeding. (Chapter 8 looks at whether unanticipated effects are more or less likely or more or less dangerous in molecular breeding than in traditional breeding.) Given the high rate of untoward effects, it is not surprising that scientists continue to search for more efficient and dependable methods of breeding crops that will reduce the proportion of unanticipated outcomes.

Chapters 8 and 9 will address stage 6, the final stage, where the transgenic crop is evaluated for its agronomic properties and its health and environmental safety.

New Developments in Molecular Breeding

A second wave of biotechnology emerged within ten years after the first commercial products were introduced by the transgenic methods previously described. By 2015, the European Commission was considering seven new genetic engineering techniques to determine whether they would be covered by existing European Union laws and whether they would be classified under the definition of GMO.

The second generation of molecular plant breeding techniques has been said to have one major benefit over transgenesis: they are more highly targeted. But they remain different from both conventional breeding and the first generation of agricultural biotechnology. The new plant breeding techniques are said to be designed "to produce improved crop varieties that are difficult to obtain through traditional breeding methods ... but the resulting end products do not contain any foreign genes."[9] This view suggests that the new techniques only edit existing genes in the plant cells; that if they transfer a transgene, they are able to ensure that no other foreign genes are introduced; and that if foreign genes are introduced, they can be bred out. Five of these techniques in molecular breeding include gene editing, oligonucleotide-directed mutagenesis (ODM), cisgenesis and intragenesis, RNA-dependent DNA methylation (RdDM), and synthetic DNA.

Gene Editing

Methods classified as gene editing afford plant breeders the ability to alter the sequence of DNA in a plant cell at predetermined sites in the genome. The method can make small or large insertions or deletions. Two techniques under the category of gene editing are zinc finger nuclease technology and CRISPR/Cas9.

Zinc finger nucleases (ZFNs) are a class of genetically engineered DNA-binding proteins, which are used in the targeted editing of an organism's genome. The ZFNs contain multiple finger-like protrusions that make contact with their target molecule. ZFNs can create breaks in DNA at a specified location. They have been called custom-designed molecular scissors.

Currently, the most widely used gene editing method is called *CRISPR/Cas9* (clustered regularly interspaced short palindromic repeats). Originally discovered in bacteria, which use CRISPR to protect themselves from invading viruses, the DNA-RNA complex consists of DNA repeats, proteins, and RNA molecules that can be tailored to recognize and edit a DNA site in an organism. The bacterial system was the model that was adapted for eukaryotic cells. Thus, CRISPR provides genetically engineered site-specific nucleases that can insert, delete, or replace DNA sequences. It has become an increasingly popular method for gene deletion and gene stacking for transgenic or cisgenic breeding.[10] The optimism that CRISPR/Cas9 or its close relatives will be the future of plant breeding is expressed in this prediction in *Frontiers in Plant Science*: "The CRISPR-Cas9 holds a very promising future in making designer plants by taking only the gene of interest from a wild type species and the gene is then directly interpolated at a precise location, which in turn opens up many avenues for plant breeders for making designer plants."[11] In other words, CRISPR was adapted from its bacterial origins, which protected the bacteria from invading viruses to a form that both edits and replaces genes.

Oligonucleotide-Directed Mutagenesis (ODM)

Oligonucleotide-directed mutagenesis (ODM) is also called *site-directed mutagenesis* or *site-specific mutagenesis*.[12] First, scientists identify a target gene, and then they construct a short DNA primer of single-stranded nucleic acid, almost identical to the target gene except for one to four nucleotides (which is the mutation). The synthetic primer is introduced into the plant cell and creates a mismatch when it binds to the target gene. The cell, into which ODM is introduced, sometimes preserves the sequence of the introduced oligonucleotide that is replacing the original sequence. Plant cells that have had the ODM introduced are sequenced to determine whether they contain the desired mutation. This type of mutagenesis is considered far more specific and precise than mutagenesis involving radiation or chemicals targeted to plant cells.[13]

Cisgenesis and Intragenesis

Cisgenesis and *intragenesis* are variants of transgenic methods, but these techniques differ in the source of the genes used for transfer. *Cisgenesis* describes the transfer of alleles or genes from a crossable species into a recipient plant cell.[14] The transferred cisgenes, as they are called, code for a trait from the targeted crop species or from a sexually compatible donor. Because the plants are sexually compatible, the gene or alleles transferred are part of the conventional breeder's gene pool. Cisgenesis, which also maintains the regulatory sequences that lie adjacent to it in the donor crop, is a faster method of doing traditional breeding than making many crosses between sexually compatible plants. With *intragenesis*, the inserted sequences (alleles or genes), including the promoters and the termination sequences, may come from one or more closely related species. This method uses a hybrid DNA complex for transfer to a host, albeit from sexually compatible plants.

RNA-Dependent DNA Methylation (RdDM)

RNA-dependent DNA methylation (RdDM) is a method for changing the trait of a plant without altering its DNA. This form of epigenetics falls into the general category of RNA interference. Small double-stranded RNA molecules can silence a specific gene by closing the switch that allows the gene to express its protein product and its function. The RNA molecules direct the cell to add methyl groups to specific nucleotides along a DNA sequence. The methyl groups can silence (switch off) a gene. Certain traits (such as delayed fruit ripening, enhanced nutrients, or reduced toxins) can be produced or eliminated with RdDM. The question remains of whether altering methylation is heritable. If not, then it would not be breeding by any conventional meaning.

Synthetic DNA

Synthetic DNA is created in a laboratory and transplanted into plant cells. It is sometimes called *xenogenic*. These DNA sequences cannot be found in any living organism, so conventional breeding could never be used to acquire the genetic transfer.

Proposals for Classifying Techniques

When the new-generation molecular breeding techniques were introduced, Kaare M. Nielsen, a Norwegian professor of medicine, noted a conceptual confusion in the classification of the existing techniques.[15] He believed that the proper way to distinguish transgenic techniques was based on the genetic distance of the DNA that could be transferred between species to plants. To address the confusion in public attitudes and regulations, he proposed five categories for organisms designated as transgenic or genetically modified—intragenic (transfers among plants in the same species), famigenic (transfers within species of the same family), linegenic (transfers among species in the same lineage), transgenic (transfers among unrelated species), and xenogenic (transfers from synthetic-laboratory designed genes and a plant host). Under this classification, only intragenic and famigenic transfers can be accomplished by conventional breeding. Nielsen commented that "Current approaches to gene technology-assisted breeding have been called 'brute force' in their use of distantly related genes with little consideration for the multiple evolutionary changes that have occurred in the biochemical networks separating species."[16]

Other criteria for distinguishing GMOs from non-GMOs have also been cited in the literature. Felix Walter and Holga Puchta argue that a GMO is a plant that can be discriminated from a *natural variant* (plants that carry an induced mutation of one or a few changed nucleotides) without a transgene inserted and that a non-GMO is a plant that cannot be discriminated from a natural variant. Classical mutagenesis by chemicals and radiation has thus far been classified under traditional breeding. The authors argue that using CRISPR to create mutational changes would not qualify as a GMO crop.[17] However, this view is not universally held and would not be considered a way to circumvent GMO regulatory oversight.

The next chapter examines viewpoints about the differences and similarities between traditional and molecular breeding. Of particular concern is whether molecular breeding requires greater attention to untoward plant products and whether methods of breeding should be considered in the evaluation of risks.

3 Differences between Traditional and Molecular Breeding and Their Significance for Evaluating Crops

The controversy over GMOs is premised to some degree on the idea that they are constructed differently from traditional crops (such as by methods of hybridization or mutagenesis) and that these differences are relevant to the assessment of risk or quality. Both in the United States and Europe, some level of government oversight for new commercial crops is based on a distinction between traditional breeding and molecular breeding. Very simply, by *traditional breeding* I mean all methods of crossing among sexually compatible crops, mechanical methods of hybridization (including embryo rescue), and mutagenesis by chemicals or radiation (see chapter 1). *Molecular breeding* involves the use of recombinant DNA, CRISPR, or other forms of transgene engineering that involve the transfer foreign genes into a plant by laboratory methods. The idea that the method of breeding a crop should factor into how we assess its risks and benefits or how it should be regulated has drawn a lot of criticism. The schism has sometimes been framed as product-based versus process-based regulation.[1]

From the product-based perspective, the safety of a product should not be affected by the method used to assemble or construct it.[2] What should matter are the properties that it possesses after it is constructed. A plant cell can undergo mutations from chemicals, radiation, or gene editing by CRISPR. According to the product-based perspective, the mutations themselves may result in a safe or unsafe plant, but the methods of creating them do not. Based on this idea, "the United States had adopted a regulatory stance toward agricultural biotechnology that declined to single out GM organisms for enhanced scrutiny based solely on their method of production."[3] This product-based approach has been reinforced by several reports from the National Academies of Sciences, Engineering, and Medicine (NASEM).

From the process-based perspective, traditional breeding has evolved over a long period of time. Breeders have had opportunities to understand what can go wrong in new crop development and to become alerted to a toxic crop before it is put on the market. There is no precedent for the speed at which molecular breeding has altered the germ plasm of crops. According to the process-based perspective, the possibilities of mixing genetic material from widely diverse organisms (plants, animals, fish, and bacteria) are very likely to create a higher number of adverse effects. Thus, it is argued, the process of molecular breeding should be looked at as a special case.

Europe and the United States have adopted different perspectives on these two approaches to risk-assessment regulation. The European community has chosen a process-based approach, and the United States has chosen a product-based approach. Unlike the American regulators, the Europeans largely believe that, compared with traditional breeding, molecular breeding is likelier to introduce unique, unanticipated risks and that because of the methods used to create new crops, its products should be evaluated as a special class. Under European Union rules, all genetically engineered (GE) crops have to undergo risk assessment, and all products made from those crops must be labeled.

In its 1977 white paper titled *Research with Recombinant DNA: An Academy Forum, March 7–9, 1977*, the National Academy of Sciences (NAS) concluded that there is no evidence that unique hazards exist either in the use of rDNA techniques or in the movement of genes between unrelated organisms.[4] According to the NAS, the risks associated with introducing rDNA-engineered organisms are the same in kind as those associated with introducing unmodified organisms and organisms modified by other methods. By this view, assessment of the risks of introducing rDNA-engineered organisms into the environment should be based on the nature of the host organism, the genes transferred, and the environment into which the GMO is released and not on the method by which it was produced.[5] In its most recent and comprehensive report on GE crops, NASEM writes: "While recognizing the inherent difficulty of detecting subtle or long-term effects in health or the environment, the study committee found no substantiated evidence of a difference in risks to human health between currently commercialized genetically engineered (GE) crops and conventionally bred crops."[6]

The first issue addressed here is how scientists characterize the differences between traditional and molecular crop breeding. Then the question of whether the differences are relevant to assessing the health or environmental risks from the bioengineered crops is examined.

Wendy Harwood of the John Innes Centre in Norwich in the United Kingdom provides a useful framework for distinguishing between traditional and molecular breeding by classifying four sources of genetic variation for crop improvement. First, the primary gene pool represents the genes from the same and closely related species. These genes can be moved around by natural crosses. Second, the secondary gene pool consists of the genes from more distant species, where natural crosses are difficult but can be achieved through embryo culture methods. Third, the tertiary gene pool is only marginally sexually compatible with the plant of interest. Natural and cell culture crosses are usually not successful, but they can be done. Fourth, the quaternary gene pool consists of all organisms, including animals and microbes. Gene transfer can be done only by genetic engineering.

We have seen that, through human intervention, both traditional and molecular breeding methods exchange genes and other DNA across different species. These processes extend beyond the natural sexual reproduction of plants. In traditional breeding, clusters of genes are exchanged to an extent that depends on their genetic linkage or proximity on the chromosomes, whereas in molecular breeding, specific genes are transferred. In the former case, breeding is a hit-or-miss situation where the breeder cannot control how many and which genes are transferred. When a desired phenotype is selected, it comes with a cluster of other genes whose functions may not be understood. According to the National Academy of Sciences, when traditional breeders cross two sexually reproducing plants, tens of thousands of genes are mixed: "The major differences between traditional breeding and molecular biological methods of gene transfer lie neither in goals or processes, but rather in speed, precision, reliability and scope."[7] The NAS notes that molecular breeders move one gene at a time, whereas traditional breeders have to undertake many crosses before they observe the desired recombination of genes. There is no dispute about this characterization of the differences between the two breeding processes.[8]

There are disagreements about whether molecular breeding is a qualitatively different process—that is, whether molecular breeding radically transforms food production or whether it is a gradual extension of prior methods. Stephen P. Moose and Rita H. Mumm see ancient breeding as a form of biotechnology: "Prehistoric selection for visible phenotypes that facilitated harvest and increased productivity led to the domestication of the first crop varieties and can be considered the earliest examples of bio-technology."[9] Michael K. Hansen, senior staff scientist at Consumers Union, argues that genetic engineering is not an extension of conventional plant breeding but represents a "quantum leap" in the transformation of plants.[10]

Most commentators acknowledge that the essence of breeding under biotechnology involves cutting and splicing genes (*recombinant DNA*) to create genetic exchanges and that traditional breeding capitalizes primarily on reproduction between sexually compatible species to create genetic vari-ability from which to select desirable phenotypes.

The concept of wide crosses in traditional breeding is discussed in chap-ter 2. It is extremely rare that breeders can cross species from two distinct families or any two levels in the taxa of biological classification. This does not mean that nature does not at times exchange DNA across widely sepa-rated taxa. As Nina Federoff and Nancy Marie Brown note, "Even crossing kingdoms to put a bacterial gene into a plant is not new: Agrobacterium has done it for millennia."[11]

Bacteria and viruses that have evolved to infect plants and deposit some of their DNA into the plant's chromosome or in its cytoplasm through cir-cular plasmids do not create genetic variability for sexual reproduction of plants. However, neither natural nor human-activated traditional breed-ing can create crosses between higher taxonomic categories above genus (such as between kingdoms and orders) (see chapter 1, figure 1.1). Whether molecular breeding is qualitatively unique from traditional breeding meth-ods is not an empirical question but rather depends on the criteria that are used for assessing uniqueness and the technology available for measuring "qualitative uniqueness." For scientists who claim that molecular breed-ing is qualitatively distinct from traditional breeding, the next question is, What difference does that make for the safety and quality of the plant, and why should consumers be interested?

Michael K. Hansen of Consumers Union is among those scientists who believe that because molecular breeding draws DNA from varied sources across wide biological taxa, it will experience greater unpredictability. This unpredictability will include the safety of the product: "Because conven-tional breeding, including hybridization and wide crosses, permits the movement of only an extremely tiny fraction of all the genetic material that is available in nature, and only allows mixing and recombination of genetic material between species that share a recent evolutionary history of interacting together, one would expect that the products of conventional breeding would be more stable and predictable."[12]

It is generally recognized that the fungibility of genes through recombinant DNA escapes any barriers that might exist to natural gene exchange between biological organisms. These barriers include organisms that are generally not in contact with one another, that are reproductively incompatible, or that are not part of a genetic exchange regime in nature.

But the fact that genetic exchanges are novel and unpredictable does not necessarily make them dangerous or ineffective. Molecular breeding must work at some level to make commercially successful products. If the desired phenotype is herbicide resistance, then the seed manufacturer will have to demonstrate its efficacy for that property. Others have asked what happens when the transgene construct (marker gene, promoter, transgene, and terminator sequence) induces other properties. Is there any reason to believe that molecular breeding is likely to induce more variable or more harmful properties than traditional breeding? Will the introduction of a single, highly specified gene construct create more unexpected properties than clusters of genes combined from compatible crosses?

There are no simple answers to these questions. Without laboratory experiments or other methods of risk assessment, there can be only speculation or hypotheses. Cases of unexpected outcomes have been observed from both transgenic experiments and traditional breeding.[13] Biological systems are always defying predictable outcomes.

Yet generalizations from past experience are commonly heard, including the claim that "the many thousands of plants that had been made using these methods had not revealed unexpected hazards."[14] In one study that looked at the effect the transgene has, Maria Montero, Anna Coll, Anna Nadal, Joaquima Messeguer, and Maria Pia state that "around 35% of the unintended effects could be attributed to the process used to produce GM plants."[15]

Does nature have any protective methods for avoiding untoward genetic exchange that can be overridden by human intervention? An answer to this question may be found in the DNA constructs created for the transgenes and in the unnatural methods of delivery. Scientists have offered different reasons that a genetically modified crop is distinct from traditional breeding, which direct outcomes of the molecular breeding process.[16] First, the transgene insert may occur in the middle of a gene and may disrupt the genetic code of a recipient organism. This is referred to as the *pleiotropic effects* of integrated DNA on the host plant genome.[17] In any of the methods for delivering a gene construct previously discussed, it is generally understood that the precise location of the transgene construct entering the chromosome is not predictable and that multiple copies of the transgene construct may be inserted on the chromosome. Some scientists assert that the method of transgenesis is "imprecise, uncontrollable and unpredictable" and that "This lack of precision, control, and predictability means that the genetic engineering process can, and almost always does, result in unintended effects."[18] The second wave of gene engineering applying CRISPR gene editing may reduce the imprecision and the unintended effects, but no one believes it will eliminate them.

Second, the inserted gene is very likely to disturb the action of neighboring genes. Transgene sequences are found to influence distantly located genes as far as 8 kilobases from the insertion site.[19] The reason offered for this assertion is that a special promoter used in the transgene construct is designed exclusively for aiding foreign genes to be correctly expressed into the target plants. As previously mentioned, the cauliflower mosaic virus (CaMV) promoter can affect other genes in the plant chromosome: "Since the CaMV 35S is so strong, not only can it affect the introduced transgenes; it can also affect genes (either turn them 'on' or turn them 'off') thousands of base pairs upstream and downstream from the insertion site on a given chromosome and even affect behavior of genes on other chromosomes. Consequently, depending on the insertion site, a gene that codes for a toxin could be turned 'on,' leading to production of that toxin."[20]

Third, scientists argue that the inserted gene can produce a new protein that may be alien to the recipient organism. This is based on the idea of posttranslational modification, where the new cellular environment for a foreign gene can alter the protein structure that is synthesized by the transgenic cell from its original form found in the source (or donor) organism.

Traditional breeding involves crosses from plants that have shared a large gene pool. Even in wide crosses there is sexual compatibility. Breeders are rearranging and exchanging genes in the preexisting gene pool rather than adding and creating totally new genes. Thus, traditional breeding involves the purposeful reshuffling of genes within sexually compatible species. Why are the chances of disturbing the host genome greater with foreign genes than with reshuffling genes in a closely related species? Some new methods of biotechnology respond to the argument about foreign genes being added (or inserted) to a plant genome. The term *cisgenesis* has been coined to describe an approach to plant breeding that uses only the existing genetic resources of the plant and related species. Hongwei Hou, Neslihan Atlihanm, and Zhen-Xiang Lu begin their journal article by discussing conventional plant breeding: "Traditional plant breeding uses crossing, mutagenesis and somatic hybridization for genome modification to improve crop traits," with a goal of introducing "beneficial alleles from crossable species."[21] For transgenesis in molecular breeding, the authors claim, the sites of insertion are random and may have unpredictable side effects. They define a cisgenic plant as "a crop plant that has been genetically modified with one or more genes from a crossable donor plant."[22] One of the benefits of cisgenesis is that it avoids "linkage drag" where the transferred sequence is linked to unwanted sequences that may be transferred in traditional breeding due to their proximity to the target gene. They distinguish cisgenesis from transgenesis, although both use gene splicing and the tools of molecular genetics: "Although both transgenesis and cisgenesis use the same genetic modification techniques to introduce gene(s) into a plant, cisgenesis introduce[s] only genes of interest from the plant itself or from a crossable species, and these genes could also be transferred by traditional breeding techniques. Therefore, cisgenesis is not any different from traditional breeding or that which occurs in nature."[23]

In some ways, this technique is closer to traditional breeding because it operates within a gene pool of sexually compatible plants. But it also may include promoters and markers, which are not found in conventional plant breeding. Cisgenesis may also be combined with new gene editing technologies, like CRISPR/Cas9, which are specific to a site and known to reduce the uncertainties of random insertions.

For those who share the views of Federoff and Brown in *Mendel in the Kitchen*, too much has been made of the distinction between traditional breeding and molecular breeding. In the former case, genes are altered through mutagenesis or cell fusion in order to create new alleles with favorable crop properties. Cisgenesis is just another technique that blurs the distinction between traditional and molecular breeding. It uses molecular techniques (such as the cutting and splicing of genes) but stays within a conventionally acceptable gene pool. This raises the question of how important and acceptable the phylogenic proximity of the plant genes is in breeding to the safety and quality of the product.

In the risk assessment of GMOs, the assumption has always been that greater caution is placed in species' genetic material exchanges that have never occurred or that are unlikely to occur in nature. This was one of the precautionary consensus positions that came out of the 1975 International Congress on Recombinant DNA Molecules at the Asilomar Conference Center in Pacific Grove, California, which assessed the risks of gene splicing experiments. A coliform bacterium that was transformed with a tumor virus gene it had never seen was considered to be a risky experiment.[24]

When genetic exchanges have occurred naturally, dangerous outcomes to humans are more likely detectable by breeders or farmers because they have been expressed in nature and thus are not unexpected.

The introduction of gene editing technology to plant breeding is another path to blurring boundaries. The new tools can alter or silence gene function or expression through mutations in the plant's genome without introducing foreign DNA. Could the mutation have occurred naturally? Is that the core principle on whether the product of gene editing should be regulated? The 2001 EU directive on biotechnology does not address these questions.[25]

CRISPR/Cas9 also can be used for introducing new genes into the "editing process." But the precision of the location eliminates some of the uncertainties of the viral vectors and gene gun where the insertions cannot be controlled in the plant genome. Governments are in discussion over whether gene editing changes the regulatory landscape of transgenic plants.

There is a consensus within the scientific community that molecular breeding widens the range of possibilities for developing crops. There remain some disagreements over whether the frequency and seriousness of untoward and unexpected outcomes are increased with molecular breeding over traditional breeding, although as is shown in later chapters, that notion is put to rest by some prominent scientific groups. Mechanical methods of hybridization, embryo rescue, and mutagenesis in traditional breeding and cisgenesis in molecular breeding blur the boundaries between the two forms of crop production. Even so, the outliers of sexual versus nonsexual (recombinant DNA) crop production remain qualitatively distinct. Natural scientists are inclined to search for definitive, empirical answers to the question of whether there is a clear or unclear demarcation between

GMOs and non-GMOs. Social scientists are more likely to draw cultural explanations for dividing lines between traditional and molecular breeding. Alonzo Plough and I looked at the cultural and technical rationality of risk and found that the former is less reductionist, less beholden to the prestige principle, and more aligned with political culture than the latter is.[26] As an example, political scientist Hannes Stephan at the University of Stirling in the United Kingdom has noted that "In Europe, agbiotech has come to represent a 'sounding board' for contemporary anxieties about modernity, globalization, and the decline of national identity. In various combinations, these concerns give rise to a potent moral critique of the 'unnaturalness' of GMOs, which often crowds out utilitarian risk/benefit evaluations."[27] Chapter 8 continues this discussion by focusing on the risk of GMOs.

4 Early Products in Agricultural Biotechnology

Two agricultural products that were in development in the 1980s gave the public its first glimpse of biotechnology's early contributions to industrial farming. First, a genetically modified tomato with the trade name of Flavr Savr was promoted as a vine-ripened tomato that was less perishable and better tasting than its unmodified run-of-the-mill parental variety. Consumers were offered a new product and a new brand—the first whole-food product developed through biotechnology. The second product was a genetically engineered soil bacterium commonly called *ice-minus bacteria*, which was designed to prevent frost damage to crops.

The Flavr Savr tomato had a short commercial life as a marketed and labeled GE tomato and then as tomato puree. The GMO soil bacterium, which was designed and field tested to prevent frost damage in crops, was never commercialized as an agricultural product but nevertheless prompted considerable regulatory activity and public debate.

These two products, even in their failure, afforded the new agricultural biotechnology industry insights into the kinds of societal, market, and regulatory responses to genetically engineered products in agriculture that they would face in subsequent years. They were the product pioneers of a new biotechnology agricultural sector.

The Flavr Savr Tomato

Calgene was a West Coast biotechnology company that sought to apply new advances in molecular genetics to crop development with a focus on tomatoes, rape seed (canola oil), and cotton. The company was started in a garage in Davis, California, by a group of University of California biology professors led by Ray Valentine. Its most notable product was a slow-ripening and long-shelf-life tomato.

In 1987, Calgene researchers began investigating a tomato enzyme called *polygalacturonase* (PG) that dissolves the tomato's cell wall, which consists of a complex set of polysaccharides called *pectins*. The breakdown of pectins, a key component of most terrestrial plants, is associated with the softening of ripe tomatoes. The role of PG in tomato softening during the ripening process was reported in 1982.[1] The scientists questioned whether the expression of PG could be controlled in order to delay the rate at which ripening takes place.

Tomatoes were typically picked green from the vine so that they could be transported sometimes thousands of miles before they overripened and became unappealing to consumers. When they were ready to enter consumer markets, tomatoes were sprayed with a chemical that speeded up the ripening process and turned the tomatoes red.

Earlier, in 1978, scientists had discovered a new method of controlling protein synthesis called *antisense technology*. The basic process of protein synthesis consists of three stages: (1) DNA is copied into (2) *messenger RNA* (mRNA) (which is a single-stranded copy of the gene) by a process called *transcription*, where mRNA, with the help of transfer RNA (tRNA), serves as a template for the cell's ribosomes to (3) synthesize polypeptide chains that form a protein, a process called *translation*. Briefly, the whole process, called the *central dogma*, is described as DNA → RNA → Protein (figure 4.1).

The antisense method involves a molecule designed to bind to and interfere with a specific mRNA construct so that the protein synthesis is disrupted. In the case of the tomato, scientists were interested in disrupting the PG enzyme (protein), which is responsible for softening fruit. By reducing PG enzyme expression, the tomato could be allowed to ripen on the vine without softening. As previously noted, commercial tomatoes are harvested at the mature green stage. They are shipped long distance and ripened by treating them with ethylene for 12 to 18 hours at 20 degrees Celsius. Shipping tomatoes that are picked green, however, prevents the development of natural flavors and results in a tasteless tomato.

For the Flavr Savr tomato, Calgene scientists inserted an additional copy of the PG encoding gene but in a different orientation than the indigenous gene, flipping it upside down and backward.[2] The mirror image of the PG

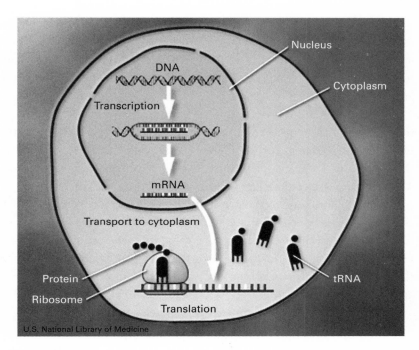

Figure 4.1
A canonical description of DNA transcription and translation in cells.
Source: U.S. National Library of Medicine.

gene codes for an antisense mRNA that is complementary and binds to the sense mRNA. This double-stranded mRNA blocks the expression of the correct mRNA. The insertion of the PG antisense gene essentially shuts off the PG enzyme and inhibits the ripening process.

The level of PG produced in the Flavr Savr tomato is as little as or less than 1 percent of the unmodified parental line. In other words, introducing a transgene in the plant that is a mirror image of a gene will code for an antisense mRNA, which blocks the expression of the correct mRNA and therefore when applied to PG reduces the amount of pectin-degrading enzyme.

With less PG produced, the tomato stays ripe and firm for a longer period of time. It can be picked red from the vine, which yields a tastier product than picking and shipping it green and then exposing it to treatment with the chemical ethylene.

The Calgene scientists produced a tomato that was alleged to be superior to the non-GM tomato in several ways. It had a longer shelf life, was promoted as tastier and of better quality, possessed equivalent nutrient content, and had higher viscosity, which was more desirable for processed juice and tomato paste.

A critical part of this process involves the creation of the DNA construct that includes the marker gene, the promoter, the transgene itself, and the termination sequence. For the marker gene, scientists used a gene from the bacterium *Escherichia coli* called kan(r), which encodes the protein APH(3')II that makes the plant cell resistant to the antibiotic kanamycin. Applying the antibiotic to the cultured cells allows scientists to tell which cells have the transgene of interest (only the ones that survive the antibiotic selection pressure). The promoter gene came from the cauliflower mosaic virus (CAMV 35s). The construct was inserted into a plasmid and through the vector *Agrobacterium tumefaciens* was inserted into tomato cells in a growth medium.

Calgene had no legal requirement to obtain regulatory approval for developing and marketing its genetically modified tomato. The U.S. Department of Agriculture's Coordinated Framework for Regulation of Biotechnology (1986) set the roles of three agencies—the U.S. Food and Drug Administration (FDA), U.S. Department of Agriculture (USDA), and U.S. Environmental Protection Agency (EPA). The USDA's oversight was over plant pests. Given that the genetically modified tomato was the first U.S. GM product commercialized, there was a heightened attentiveness to the potential risks. In October 1992, the USDA ruled that the PG antisense tomato lines were not plant pests and therefore did not require permits for field testing or transport. In 1995, the USDA added two additional genetically engineered lines to the determination of nonregulated status for Calgene's Flavr Savr tomatoes.[3]

Belinda Martineau, a Calgene scientist who wrote *First Fruit: The Creation of the Flavr Savr Tomato and the Birth of Biotech Food*, discusses the types of considerations that the company faced in doing a risk assessment of its product: "The concern over unintended changes centered on the fact that we had no control over where our genes went into a plant's DNA. They could be integrated harmlessly or smack dab in the middle of an existing gene, disrupting that gene's function. If that disrupted gene was involved in the formation of an important vitamin, for example, levels of that particular vitamin could be dramatically reduced."[4]

Calgene requested a consultation from the FDA on the Flavr Savr tomato, a decision that it made in order to gain consumer approval for the product. The company submitted safety data to the FDA on the use of the kan gene as a marker for its tomato. One concern was whether after the gene and its expressed protein were in the tomato, could they find their way into bacteria in the human gut. If that were to occur and the bacteria were to become pathogenic, they would be resistant to the therapeutic use of the antibiotic kanamycin. The FDA and a number of commentators considered the possibility to be remote. Calgene's petition for the kan gene was approved by the FDA in May 1994 after the company presented various worst-case scenarios.[5] According to Nina Federoff and Nancy Marie Brown, "When Calgene brought it to the FDA in 1992, the tomato was subjected to $2 million [or $3.4 million in 2016 dollars] worth of testing done by Calgene. ... In a public meeting the FDA scientists brought the results of their extensive and sophisticated chemical analysis to a panel of external advisors."[6]

Prior to entering the market, Calgene sponsored a number of studies to determine whether the antisense gene complex affected any other parts of the tomato besides ripening. Matthew G. Kramer and Keith Redenbaugh note "that the expression of the antisense PG gene affects only the composition of pectin in the fruit. ... The gene has no effect on levels of vitamins and nutrients, on production of potential toxins (tomatine), on taste, on non-pectin related processing traits, on horticultural traits (growth form, time to flowering, time to fruit set, etc.), fruit pH and acidity, and fruit color and size."[7]

Calgene had engaged with the FDA for five years seeking approval for its antisense tomato, during which time it undertook laboratory experiments, conducted field tests, and produced analytical papers. Finally, on May 18, 1994, the FDA concluded that the Flavr Savr tomato had not been significantly altered compared to nontransgenic varieties. Roger Salquist, CEO of Calgene, commented on the long-awaited approval: "The event of last week may have been the single most important step in establishing this new industry."[8]

Beginning in May 1994, the Flavr Savr tomato was marketed in California and Illinois. It was sold under the MacGregor brand and labeled as a genetically engineered product, the first voluntarily labeled GM crop. At the time, there was growing competition for tastier, farm-fresh tomatoes, including hydroponically grown varieties and others green harvested for better transport. In that stiff competition, the success of Calgene's tomato depended on how well it performed during shipment.

Belinda Martineau describes her first experience with the genetically modified tomato compared with its nonmodified strain: "Both groups of tomatoes were left at room temperature and observed over time. After 3 to 4 weeks the nongenetically engineered tomatoes were noticeably shriveled and rotting, while the Flavr Savr™ tomatoes looked and felt essentially no different than first picked."[9]

But after the tomatoes emerged from the greenhouse and entered the distribution system, problems emerged. One of Calgene's scientists reported on the results of the company's transport management: "Contrary to the company's early expectations, the Flavr Savr gene could not keep vine-ripened fruit firm enough to be packed and transported like green tomatoes."[10] High production costs, poor growing conditions in Florida, steep competition, and company debt led to the demise of the Flavr Savr tomato, despite the fact that the tomato's genetically designed property of slow softening was successful (although some noted that the taste was not superior). Calgene spent more money getting the tomatoes to market in good shape than it charged in the grocery store—an unsustainable business plan. By 1996, Flavr Savr tomatoes were removed from the consumer market in the United States. They were not approved for sale in the European markets, although a puree made from genetically engineered tomatoes had a short success in the United Kingdom during 1996 to 1999. Public attitudes had turned against GE products, which dramatically affected the market share of GE tomato puree and its adoption in major UK food markets.

Molecular breeding involving antisense technology was a small step in the move toward genetically engineered crops. No new genes were introduced into the tomatoes: the gene for one enzyme was removed and inverted. Yet questions were being asked. Would there be any other effects from the insertion of the antisense PG gene? Although the FDA did not require those questions to be answered, Calgene needed to ensure that it would not face any unexpected outcomes before it took the tomato to market and that its rigorous premarketing review would make the product acceptable to skeptical stakeholder groups. Ironically, the Flavr Savr tomato's short commercial life precluded it from being tested in the national consumer market.

A Microbe That Protected Crops from Frost

In areas of the country prone to spring frosts before the fruit and vegetable harvest, like the orange groves of south Florida or the strawberry fields of northern California, freezing temperatures can inflict significant damage to crops. Frost is one of nature's unpredictable pests. To save their crops from frost damage, farmers have used a variety of methods, some of which can be environmentally detrimental, labor intensive, and costly. All these methods have the same goal—to keep the temperature of the frost-sensitive plants above the temperature at which frost forms. These methods have included using wind machines that mix warmer air with cooler air, burning fuel that heats the air around the crops, watering the soil to release heat to the frost-forming droplets coating the leaves of the plants, and creating artificially generated fog that shields the plants from heat loss. One of the methods is akin to fighting fire with fire. When water droplets on the plants are beginning to freeze, the farmer sprays water on the plant: "The latent heat of fusion, released when water freezes to form ice, maintains the ice-water mixture on leaves at 0°C. This mixture will remain at 0°C as long as sufficient water is continuously available to freeze."[11] Ironically, the crops are destroyed not by the temperature per se but by the ice crystals that interfere with the plant's metabolism.

In 1974, a scientific team at the University of Wyoming published a paper titled "Ice Nucleation Induced by *Pseudomonas syringae*" in which they concluded that the soil microorganism may play a role in the production of ice nucleation that occurs in nature.[12] Building on that work, a group of scientists at the University of California at Berkeley came up with a novel approach for mitigating frost damage in crops. *P. syringae*, which is ubiquitous in agricultural fields, has a protein complex in its outer membrane that provides the nucleation site for ice crystallization. The Berkeley scientists discovered that *P. syringae* had one particular protein that seeded ice formation. In 1982, Steven E. Lindow and his colleagues observed that higher concentrations of *P. syringae* on leaf surfaces were associated with warmer freezing temperatures. These bacteria were called *ice nucleation active* (INA): "Water in plant tissues can be supercooled to –5°C, without harming the plants, yet when the ice nucleating bacteria (INA+) are present, ice formation can begin at about –1°C."[13] The four degrees can be a significant difference in frost damages.

Two approaches were initially taken to apply the knowledge of ice nucleation for frost protection. The first involved eliminating INA+ bacteria on plants by locating a virus that could kill the bacteria. With no INA+ bacteria, there will be no ice nucleation. That work was supported by University Genetics Company of Norwalk, Connecticut, and the Frost Technology Corporation. But the method proved too difficult to gain results.

The second approach adopted by the Berkeley group used gene splicing methods to remove the genes responsible for the ice nucleating proteins. In other words, they created an ice-minus form of *P. syringae*. Their idea was to spray high concentrations of ice minus on the crops and thereby displace the INA+ bacteria. Lindow worked with Advanced Genetic Sciences to bring the technology into a commercial application.

In 1982, Lindow and his colleagues at UC Berkeley in collaboration with Advanced Genetic Sciences petitioned the Recombinant DNA Molecule Advisory Committee (RAC) of the National Institutes of Health (NIH) to field test ice minus. On June 1, 1983, NIH authorized the field test, but litigation delayed the testing of this genetically modified frost protectant for another year. Because this was the first environmental release of the genetically modified bacteria, there were immediate reactions from environmental groups that questioned the safety of the product. Ice minus was classified as a pesticide by the EPA on the basis that frost was considered a plant pest. This designation played a major regulatory role in the first field tests and gave regulatory oversight to the EPA. The EPA's role began in 1982, and it took five years of reviews and court decisions before the first field tests took place in Brentwood, California, in April 1987 under intense media attention.

The principal risk concerns that environmental activists and communities raised were the effects that the ice-minus bacteria might have on precipitation patterns. It was feared that widespread use of the bacteria could cause the ice minus to waft into the atmosphere, cause cloud droplets to form ice crystals, and alter natural rainfall patterns. In response to the concerns, Monterey County, California, passed an ordinance prohibiting field tests in the county. During the first field trial, the EPA monitored movement of the ice-minus bacteria to assess atmospheric risks. None had been identified. The now defunct Office of Technology Assessment (OTA) reported on the scenario of reducing atmospheric concentrations of ice nuclei affecting rainfall: "Seeking to assess the likelihood of this scenario, OTA commissioned two analytical studies by groups taking slightly different approaches to the problem. Both groups made assumptions to produce a worst case scenario. They both concluded it is unrealistic to expect any significant negative impact on global climatological patterns from large-scale agricultural applications of ice-minus bacteria."[14]

The engineered form of ice minus was found to occur in nature at very low frequencies. Under the laws in effect at the time, the release of the natural form of ice minus would not fall under regulation. Meanwhile, Advanced Genetic Sciences merged with the company DNA Plant Technology to form DNAP: "DNAP's selection and marketing of a nongenetically engineered organism for frost control, was an interesting overall strategy to bypass the complex rules associated with the introduction of genetically engineered organisms."[15] The new company focused on using naturally occurring ice-minus strains for developing several formulations of Frostban but later drifted from frost control to disease control.[16]

Frostban, even with naturally selected ice-minus strains, was not a profitable product. Estimates of its profitability were downplayed.[17] Ironically, it was the natural bacteria, ice plus (of which there was a plentiful supply), that became profitable for inducing snow production in ski resorts. No risks were confirmed with the use of Frostban. The case study illustrates the process of creating regulations on the spot because the EPA was operating on a regulatory framework developed for chemical pesticides and not genetically engineered biological agents.

These two bioengineered products, the Flavr Savr tomato and Frostban, are emblematic of developments to come. Improving the taste and extending the shelf life of a tomato were benefits to the consumer as well as the distributor. The next generation of genetically engineered crops displayed traits that benefited growers, such as insect and herbicide resistance. Frostban raised questions about the environmental release of bioengineered organisms, foreshadowing genetically modified salmon, crops with built-in bacterial toxin genes, or gene-drive technology, where gene-edited mosquitoes were genetically designed to resist malarial infection.

These early products taught scientists about the slow adjustment that society makes to new food technology and agricultural processes. One cannot compare the speed of society's adaptation to digital technology to that of its adaptation to food technology. These early cases also reveal differences between cultural and technological forms of risk assessment. Public skepticism toward new products or technologies is not usually resolved by a laboratory experiment. The public must see that bioengineered products have benefits over existing products and come to believe that those benefits outweigh the risks or uncertainties. These early cases suggest that people are more inclined to believe in agencies of government that are independent of the manufacturing sector and that they are not reluctant to trust scientists who play a central role in evaluating risks but who do not have a profit-making interest in the bioengineered products they evaluate for society.

5 Herbicide-Resistant Transgenic Crops

Herbicides entered U.S. agriculture after World War II. The first commercially available herbicide was 2,4 dichlorophenoxy-acetic acid (2,4-D), introduced in 1946.[1] It was developed during the war by independent research groups in the United States and the United Kingdom under conditions of wartime secrecy. Scientists were looking for chemical warfare agents (biocides) that could be used to destroy agricultural fields of Germany and Japan and create famine as an ancillary military strategy to conventional ground warfare and aerial bombing.

In August 1944, the journal *Science* published one of the earliest nonclassified studies of 2,4-D and its use on bindweed: "By the fifth day following application of the spray, the basal leaves were yellow, and at ten days the above ground parts were dry and dead."[2]

In her widely acclaimed 1962 book *Silent Spring*, Rachel Carson devoted several pages to 2,4-D, hinting at its toxicity and alerting readers to its unanticipated environmental effects: "With the widespread use of 2,4-D to control broad leaved weeds, grass weeds in particular have increasingly become a threat to corn and soybean yields."[3] Carson referred to 2,4-D as one of the most widely used herbicides of her day. The herbicide 2,4-D selectively controls broadleaf weeds but allows grasses to remain relatively unaffected.

From the farmer's perspective, "herbicide treatments are an integral part of modern agriculture because they provide cost-effective increases in agricultural productivity."[4] Food crops fare better without competition from weeds for water, light, nutrients, and space and without contaminating weed seeds. Weeds often play a role as a host or shelter for plant pathogens that affect food crops in quality and quantity. Herbicides also help protect soil conservation by supporting no-till agriculture. Tillage (the turning over of soil) contributes to erosion, the reduction of soil fertility, and the loss of top soil. At first glance, herbicide-resistant (HR) plants can be seen as contributing to no-till agriculture, which helps to sustain the soil. But as Martin Paul Krayer von Krauss, Elizabeth A. Casman, and Mitchell J. Small note, after you take into consideration "volunteer plants" (plants found growing without having been planted), tillage creeps back in: "To counter the [herbicide]-tolerant volunteer problem, farmers will have two options: use additional herbicides or revert to tilling. Experts fear that the cost of the first option will exceed the cost of the second … a large- scale reversion to tilling would be expected, and the soil conservation and agronomic benefits of no-till agriculture would be lost."[5] Tillage also creeps back in when weeds become herbicide-resistant.

The use of herbicides in the United States increased dramatically in the last quarter of the twentieth century. In 1974, 800,000 pounds of glyphosate was used in U.S. agriculture, and by 1995, its use had tripled. Nearly twenty years later in 2014, its applications grew to 250 million pounds.[6] Between 1974 and 2014, about 3.5 billion pounds of glyphosate was applied in the United States. According to the U.S. Department of Agriculture, "an overreliance on glyphosate and a reduction in the diversity of weed management practices adopted by crop producers have contributed to the evolution of glyphosate resistance in 14 weed species and biotypes in the United States."[7]

The chemical glyphosate was first synthesized by a Swiss chemist in 1950 but not developed as an herbicide. It was rediscovered in 1970 by a Monsanto chemist who recognized its herbicidal properties.[8] Glyphosate was patented and marketed in 1974 under the trade name Roundup. The herbicide won awards for the Monsanto chemist for its broad-spectrum effects that controlled both broadleaf and grass weeds and for what was then understood to be its low toxicity to animals compared to other herbicides. Prior to the development of recombinant DNA, herbicide resistance in plants was discovered by trial and error without an understanding of the mechanism that made crops tolerant to a specific herbicide. Some efforts were made to breed certain cultivars with herbicide resistance. Several canola varieties that were resistant to triazine (a family of herbicides) were released in the 1980s, but lower yields limited their commercial value. There were also some natural strains of rice that were resistant to imidazolinone. In the 1960s and 1970s, seed and herbicide manufacturers were different companies. There was no incentive for seed producers to breed herbicide-resistant crops or varieties to benefit chemical companies. That changed in the 1980s, when large chemical companies bought up seed producers.

Monsanto scientists discovered a gene for a glyphosate-insensitve form of 5-enolpyruvylshikimate-3-phosphate synthase (EPSPS), an enzyme that is critical to plant growth and that was found in some microorganisms. After the EPSPS variant is incorporated into the plant genome, the gene product confers crop resistance to glyphosate. In 1996, Monsanto introduced transgenic glyphosate-resistant soybeans into U.S. agriculture and followed it with other crops (table 5.1). The seeds soon became commercially successful, and with that success, the company's glyphosate formulation called Roundup became the leading herbicide applied in the United States.

The developers of glyphosate-resistant crops announced the agricultural benefits of the herbicide prior to its commercial release: "Glyphosate has favorable environmental features such as rapid soil inactivation and degradation to natural products, little or no toxicity to important life forms, and minimum soil mobility."[9]

Table 5.1
Herbicide-resistant genetically engineered crops approved for sale in the United States, 1994 to 2006

Crop	Glufosinate	Glyphosate	Bromoxynil	Sulfonylureas
Cotton	2004	1997	1994	
Soybean	1996	1996		
Ccanola	1995	1996	2000	
Maize	1997	1998		
Sugarbeet	1998	1999		
Rye	1999	1999		
Flax				1999
Rice	2006			
Alfalfa		2005		

Sources: Stephen O. Duke, "Taking Stock of Herbicide-Resistant Crops Ten Years after Introduction," *Pest Management Science* 61 (2005): 211–218; Stephen O. Duke and Antonio L. Cerdeira, "Transgenic Crops for Herbicide Resistance," chap. 3 in *Transgenic Crop Plants*, vol. 2, *Utilization and Biosafety*, ed. Chittaranjan Kole, Charles H. Michler, Albert G. Abbott, and Timothy C. Hall (Heidelberg: Springer, 2010), 134.

Scientists at Monsanto heralded herbicide-resistance technology as a breeding mechanism that contributes to the sustainability of world agriculture through the use of high-quality and safe herbicides.[10] Weed scientists largely supported the introduction of transgenic crops resistant to broadspectrum, nonselective herbicides. During the 1990s, the two most popular herbicides approved in conjunction with herbicide-resistant crops were glyphosate and glufosinate.

The scientific work leading to the development of glufosinate dates back to the 1970s. The herbicide contains phosphinothricin, a chemical that kills plants by blocking a plant enzyme that is critical for nitrogen metabolism and for detoxifying ammonia, a by-product of plant metabolism. Herbicide-resistant glufosinate seeds contain a bacterial gene that encodes an enzyme that detoxifies phosphonothricin and protects the crop. Like glyphosate, glufosinate is a broad-spectrum herbicide that also kills fungi and bacteria. Table 5.1 shows the dates of approval of herbicide-resistant seeds for four herbicides from 1994 to 2006.

By 2013, 93 percent of all soybean acres, 85 percent of corn acres, and 82 percent of cotton acres were of herbicide-resistant (also referred to as herbicide-tolerant or HT) varieties. Public debates over herbicide-resistant crops raised the following questions: Do the transgenic HR crops provide added value to farmers (in the form of better yields and more income)? Do HR seeds require more or less use of herbicides? Does the use of HR crops spread weed resistance to herbicides? Are HR seeds used with toxicologically safe herbicides for humans and wildlife? Do HR crops improve or at least leave undiminished the ecosystems of both farms and surrounding areas?

On the question of yields, a USDA report stated: "The evidence on the impact of HT seeds on soybean, corn and cotton yields is mixed ... several researchers found no significant difference between the yields of adopters and nonadopters of HT; some found that HT adopters had higher yields, while others found that adopters had lower yields."[11] The nonadopters may be using nontransgenic crops that are tolerant to selected herbicides. Yields have a lot to do with multiple factors. The most dependable studies require side-by-side plantings using the same herbicide with transgenic and nontransgenic crops.

In a comprehensive review of the scientific literature, Doug Gurian-Sherman, of the USDA and EPA but writing as a senior scientist for the Union of Concerned Scientists, concluded: "it does not appear that transgenic HT corn provides any consistent yield advantage over several non-transgenic herbicide systems... motivations other than increased yield are more likely to be encouraging farmers to adopt HT corn."[12] One comparative yield study by a group at the University of Nebraska found that the yields of glyphosate-tolerant cultivars were 5 percent lower than yields of the non-GMO sister cultivars. The authors surmised that the yield suppression in the GMO crops was related to the transgene or its insertion process, which raised doubts about the acclaimed yield gains from GMOs.[13]

A second contested issue is farmer return on investment. It has been widely reported that herbicide-resistant crops are associated with lower weed-control costs, especially compared to mechanical methods of weeding. Although there are some recognized cost savings in reduced weed management with HR crops, those are somewhat offset by higher prices for the seeds. The USDA study claims that there is a mixed effect on net returns. It argues that by reducing the time for weed management, farmers are able to work in other areas and even take on other jobs, which increases their net returns.[14]

Many claims and counterclaims have been made about whether HR crops reduce the use of herbicides. There is clearly a distinction between insect-resistant crops (which is discussed in chapter 7) and herbicide-resistant crops. The purpose of insect-resistant crops is to place the insect repellant in the plant rather than spraying it on the plant. But HR crops are designed to be used with herbicides. One claim is that the transgenic HR crops are more efficient and therefore require less herbicide use. Another claim is that transgenic HR crops are linked to a safer herbicide than non-transgenic crops. According to a USDA study by Jorge Fernandez-Cornejo, Seth Wechsler, Mike Livingston, and Lorraine Mitchell, "The adoption of HT [referred to as herbicide-tolerant rather than herbicide-resistant] crops has enabled farmers to substitute glyphosate for more toxic and persistent herbicides."[15]

Studies have shown that for HR cotton and soybean, herbicide use (measured in pounds per acre) declined slightly in the first years and then increased. The USDA calls the increase "modest."[16] The USDA data suggest otherwise: "Herbicide use on corn by HT adopters increased from around 1.5 pounds per planted acre in both 2001 and 2005 to more than 2.0 pounds per planted acre in 2010, whereas herbicide use by nonadopters did not change much."[17] That is a hardly modest 30 percent increase.

A suggested reason for the increase is that weeds became resistant to the primary herbicide—glyphosate. Nonadopters were more likely to diversify their management of weeds to stay on top of weed resistance. With *integrated weed management*, farmers utilize multiple strategies to control weeds. Those locked into the HR seeds increased their herbicide use in response to growing weed resistance. By 2014, seeds that were resistant to multiple herbicides with different modes of action (known as *pyramiding* or *gene stacking*) were approved. Enlist Duo seeds are resistant to glyphosate and 2,4-D.[18] This was the industry's response to the rise in weed resistance from a single herbicide.

Charles M. Benbrook, former executive director of the board of agriculture of the National Academy of Sciences, studied pesticide use in the United States during the growth of genetically modified crops. One part of his report covered herbicides and herbicide-resistant crops. He concluded that "HR crop technology has led to a 239 million kg (527 million pound) increase in herbicide use across the three major GE-HR crops, compared to what herbicide use would likely have been in the absence of HR crops. Well-documented increases in glyphosate applications per hectare of HR crop account for the majority of the 239 million kg increase."[19] His results are consistent with the finding that the spread of glyphosate-resistant weeds has led to increases in the use of glyphosate and other herbicides. Edward D. Perry, Federico Ciliberto, David A. Hennessy, and GianCarlo Moschini found that "Since 1998 the most striking trend has been an increase in the use of glyphosate."[20]

Benbrook also found that glyphosate application has risen nearly fifteen-fold since Roundup Ready crops were first introduced in 1996 and that transgenic HT crops account for about 56 percent of the global use of glyphosate.[21]

Do HR crops contribute to weed resistance? The consensus viewpoint is that the monolithic continuous uses of a single herbicide will eventually contribute to weed resistance. This was predicted in 1996 at the dawn of the revolution in transgenic crops: "The widespread use of HRCs developed for resistance to single herbicides will accelerate the selection pressure on weeds to evolve resistant biotypes."[22] And these resistant biotypes have to be able to reproduce with their fitness to survive unaffected.

This prediction is particularly relevant with development of glyphosate resistance at a global scale as shown in the number of weed species across

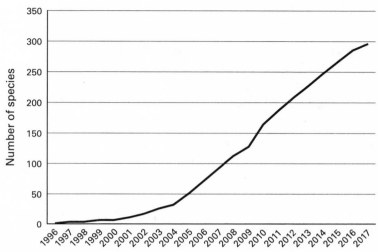

Figure 5.1
Global increases in glyphosate-resistant weeds.
Source: Weedscience.org, International Survey of Herbicide-Resistant Weeds,
http://www.weedscience.org/summary/resistbyactive.aspx.

the globe that have developed glyphosate resistance over two decades
(figure 5.1).

The next generation of HR crops is being developed with stacked resis-
tance to multiple modes of herbicide action into a single seed. As Jerry M.
Green has noted, "Combining herbicide mixtures with multiple-resistant
(MR) varieties can reduce reliance on a single MOA [mode of action]."[23]

Have HR crops been developed with the safest herbicides available? This
was certainly one of its early appeals. Because the most prevalent herbicide
in current use is glyphosate, scientists have asked whether glyphosate is
the safest broad-spectrum herbicide that could be used for HR crops and
whether its level of safety is safe enough. One way to document toxicity is
by oral LD_{50} values, which is the amount of the chemical required to give
a lethal dose to the test animal population, usually mice. Scientists at the
Institute of Food and Agricultural Sciences at the University of Florida compared the oral LD_{50} of several commonly used pesticides (the lower the lethal
dose, the higher the toxicity): paraquat ~100; triclopyr 630; 2,4 D 666; penda-
methalin 1,050; atrazine 3,090; glyphosate 4,900; imazaquin >5,000.[24] On
the LD_{50} criteria, glyphosate comes out fairly well.

Other toxicological criteria include whether the substance is a carcinogen, neurotoxin, mutagen, or endocrine disruptor. Also to be considered are the formulations for glyphosate, which includes adjuvants that amplify permeability and increase toxicity. Roundup Ultra, the glyphosate herbicide for Roundup Ready crops, contains 41% of glyphosate and 59% of other ingredients.[25] A 2009 study confirms that the adjuvants for Roundup (a trade variety of glyphosate) can sometimes be more toxic to organisms than the active ingredient: "adjuvants in Roundup formulations are not inert. Moreover, the proprietary mixtures available on the market could cause cell damage and even death around residual levels to be expected, especially in food and feed derived from R formulation-treated crops [that is, POEA]."[26] A follow-up study by Nicolas Defarge and his colleagues looked at the effects of glyphosate's active ingredient and adjuvant chemicals on human cells: "Briefly, all co-formulants inhibited aromatase and disrupted mitochondrial respiration (and membranes) at higher concentrations. APG and POEA were 15–18 times and 1200–2000 times more cytotoxic than G, respectively."[27]

There are also environmental criteria. How does the herbicide compare with other herbicides with respect to their environmental effects on nontarget plants and animal species? Comparative evaluations of herbicides for environmental effects are rarely definitive. The decline of the Monarch butterfly has spurred studies on whether herbicides are one of its causes, and glyphosate has been identified as a major suspect:[28] "Given the established dominance of glyphosate-tolerant crop plants and widespread use of glyphosate herbicide, the virtual disappearance of milkweeds from agricultural fields is inevitable. Thus, the resource base for Monarchs in the Midwest will be permanently reduced."[29]

Although glyphosate—compared in all categories to 2,4-D, atrazine, and bromoxynil—has traditionally come out as safer, that conclusion was brought into question when the International Agency for Research on Cancer reclassified glyphosate as a probable human carcinogen.[30] Recent studies have established new evidence questioning the safety of glyphosate: "Research has now established that glyphosate can persist in the environment, and therefore, assessments of the health risks associated with glyphosate are more complicated than suggested by acute toxicity data that relate primarily to accidental high-rate exposure ... chronic glyphosate exposure at low concentrations can potentially result in risks to human health."[31]

The toxicity of glyphosate-based herbicides must be understood through the different formulations of the active ingredient, some of which contain toxic adjuvants that are reputed to have a higher toxicity than pure glyphosate.[32]

The final word on the toxicology of glyphosate has no direct bearing on the safety of the herbicide-resistant-glyphosate seed. But it does raise the question about a popular seed class that is inextricably tied to the use of a particular herbicide formulation that has become increasingly suspect and the target of scores of lawsuits.

6 Disease-Resistant Transgenic Crops

The three scourges of agricultural crops are weeds, insect predators, and infectious pathogens (namely, bacteria, viruses, and fungi). Before chemical herbicides were discovered, weeding was handled by mechanical methods. And before chemical insecticides were known, farmers learned to mitigate insect pests by planting techniques such as intercropping, crop rotation, and biological controls. Today organic farmers still use these practices in lieu of chemical pesticides.

Many plants possess natural defenses against some infectious pathogens. Although plants do not have an immune system comparable to that of animals, they do have defense mechanisms against invading microbes that exhibit some similarities in their mode of action to the immune system of animals: "Plants react to pathogen attack by activating a variety of defense mechanisms that culminate in a number of physical and biochemical changes in the host plant."[1]

As in all forms of biological life, plants can detect threats to their survival and react to those threats. Plants can react to certain chemicals through receptors that are located either on their cell surface or inside the cell. After the receptors are activated by the coat proteins of a pathogen, they will issue an antiviral, antifungal, or antimicrobial response.

For example, plants naturally defend themselves against pathogens by the thickening of cell walls. If agents penetrate the walls, plants release protective proteins (pathogenesis-related proteins or phytoalexins, which are small antimicrobial chemicals that accumulate in plants as a result of infection or stress) that protect them from the immediate and future pathogen assaults. For example, scientists discovered a gene in maize that encodes an enzyme that protects it from a fungal attack by disabling the invading pathogen. In a process known as *systemic acquired resistance* (SAR), a plant can develop greater resistance to a pathogen after its first attack.[2] SAR was first observed in 1901and has been more extensively studied over the past two decades. A leaf that is not in direct contact with a pathogen can be stimulated to protect itself from a part of the plant that has been infected— a process known as *inducible microbial proteins*. In some cases, farmers can spray crops with chemicals called *plant activators* that can artificially trigger systematic acquired resistance, essentially activating the plant's protective mechanism. These chemicals often have less toxicity to animals and humans than traditional fungicides.

Like human pathogens, plant pathogens are constantly evolving and mutating to overcome crop resistance. In agriculture, infectious pathogens represent a significant economic loss and control cost to farmers. In the United States, it is estimated that plants are subjected to over fifty thousand different pathogens (fungi, viruses, bacteria, and nematodes). Any given agricultural region may be faced with between ten and fifteen serious plant diseases. From 1988 to 1990, there was a loss of $300 billion for eight major crops from all sources (pathogens, arthropods, and weeds), and about a third ($100 billion) of this loss was due to pathogens. Nonindigenous pathogens crossing national boundaries contribute significantly to the total crop loss:[3] "Even with the extensive application of pesticides, the estimated reductions in the farm-gate value of selected vegetable crops in the United States caused by diseases range from 8 to 23%, by insects 4 to 21%, and by weeds 8 to 13%. The average losses caused by diseases, insects and weeds in Canada are 15.5, 12.5 and 10.5%, respectively. They reduced return to the vegetable industry by $172.7, $138.2 and $115.2 million, respectively, in 1990."[4] On the global scale, it is estimated that 10 to 16 percent of the total harvest is lost to plant diseases each year, costing an estimated $220 billion.[5]

Papaya trees can be found in the tropical and subtropical parts of the world.[6] The papaya fruit (called the fruit of angels) was first introduced to the Hawaiian islands in the 1940s and is the second-largest fruit crop in the state. In the 1950s, a virus called papaya ringspot virus (PRSV) (the disease name came from the ringed spots of the fruits of infected trees) destroyed the papaya crop on Oahu island. As a result of the PRSV blight, the industry relocated to the Puna district of Hawaii island. For a while, the region was free of the virus, but in 1992, PRSV appeared in Puna, which was then Hawaii's major papaya growing region, resulting in a decline in production by 50 percent.[7] The virus is spread by aphids, which feed on the leaves of infected trees, carry the virus on their mouthparts, and transmit the virus to healthy trees.[8] Young seedlings die quickly and usually do not produce healthy fruit. The virus attacks mature trees that develop mosaicism, distortion of leaves, and smaller fruit with ringspots until the trees die. Efforts to remove infected trees do not prevent the spread of the virus.

In the late 1980s, prior to the development of transgenic crops, scientists built on knowledge of systemic acquired resistance and used a method for protecting crops from pathogens that is similar to the stimulating of a human immune response to a virus. Dennis D. Gonsalves and a team at Cornell University and the University of Hawaii inoculated papaya seedlings with a mild strain of PRSV. The mild strain was developed by exposing a severe strain to nitrous oxide, which is a mutagen. Gonsalves and his colleagues found that papaya trees exposed to the mild strain of PRSV were protected from the severe strain.[9] The method, which is called *cross-protection strategy*, was reported to have worked effectively in Taiwan from 1984 to 1993. However, it had limitations that included the nonavailability of stable attenuated strains, the requirement of a large-scale inoculation facility, and its failure to protect the trees from new strains in different countries.[10]

A physical method that was used to protect papaya crops against PRSV involves covering an entire orchard with netting to provide a barrier to the aphids that transmit the virus. However, this method reduces the sunlight to the trees, resulting in lower-quality fruit. Also, after the harvest, the netting is removed and burned, which results in hazardous by-products.

Modern developments in biotechnology, particularly recombinant DNA methods, led to a new approach to the PRSV virus called *pathogen-derived disease resistance*. It is based on a similar idea of the *cross-protection strategy*. Under the right circumstances, plants can be sensitized to protect themselves against invading pathogens. The plant is exposed to fragments of the invading pathogen. The transgene, which is made up from part of the pathogen's genome, is connected to a promoter like the cauliflower mosaic virus and delivered into the plant host cells by *Agrobacterium tumefaciens* or a gene gun. In the gun, gold or platinum particles are laced with the transgene cassette. After the transgene cassette enters the plant cells, the cells are activated to express RNA or proteins that disable the invading pathogen. The plant response occurs after the plant attack. The chemicals that are produced by the plant are called *inducible defenses*. This type of generalized defense response is called *RNA silencing* or *RNA interference*. The method was deemed effective and safe according to a report published in the journal *Science* in the mid-1980s: "The health and environmental issues can be alleviated or totally avoided by selecting appropriate sequences from pathogens and engineering them into safer transgenes to preempt the possibility of their expression and/or persistent presence by pathogenic proteins or RNA sequences in transgenic hosts.... RNA silencing-mediated transgenic resistance is one of the strategies that does not create pathogenic nucleotide sequences or proteins in the host and is totally human-, livestock-, and environment-friendly."[11]

A variant of this method was successful in protecting tobacco plants from the tobacco mosaic virus (TMV). A coat protein of TMV was inserted into the tobacco plants. The method is also termed *coat protein gene-mediated transgenic resistance*.[12] Two genetically-modified cultivars of papaya (Rainbow and Sunup) were deregulated by the Animal and Plant Health Inspection Service (APHIS) of USDA and the EPA in May 1998. The results of the transgenic papaya cultivars resistant to PRSV in Puna, Hawaii, were deemed highly successful.

A mild Hawaiian PRSV strain was the source of the coat protein gene for the transgene cassette that was to be inserted into the papaya cells. Scientists used the cucumber mosaic virus to obtain the promoter for the transgene. In addition, they used kanamycin-resistant markers and biolistics (gene gun) in conjunction with an embryonic tissue culture technique to develop the disease-resistant papaya. The transgenic plants were tested with controls in field experiments. The transgenic papaya line had no symptoms of the pathogen, whereas the nontransgenic plants were ruined within five months. One scientific group concluded: "The transgenic resistance conferred by the viral CP [coat protein] gene has become the most effective method to prevent papaya from infection by the noxious PRSV.... Thus, the GM papaya has saved the papaya industry in Hawaii, without any significant adverse effects to environment and human health during the application for more than a decade."[13]

What considerations were given to understanding the potential adverse effects of "inducible defenses"? The primary environmental risks considered by scientists and regulators were related to the transfer or recombination of transgenes in the crop, including the movement of the transgenic construct to nondomesticated papaya or wild relatives (by pollen or horizontal gene flow) and weediness of virus-resistant papaya. Among the human health effects, regulators were concerned about whether kanamycin resistance would spread and whether the essential vitamin composition or traces of toxins (such as alkaloids) of the transgenic cultivar were significantly changed.

The term *heteroencapsidation* refers to the process by which a virus's nucleic acid is enclosed in a capsid by the coat proteins of another virus. For GMOs that are made disease resistant, this means that the coat proteins (CP) of the pathogenic virus inserted into the plant cells will be able to encapsulate the RNA genome of another virus that infects the plant. This is particularly relevant when a plant is infected by more than one virus. Scientists have speculated on the possibility that heteroencapsidation could result in new virus epidemics. This natural mechanism has been documented in some transgenic herbaceous plants. In reviewing the risks, Marc Fuchs and Dennis Gonsalves observe that "heteroencapsidation in transgenic plants expressing virus CP genes has been of limited significance and would be expected to be negligible in regard to adverse environmental effects."[14] Thus, these authors dismiss the risks without the need for further tests or evidence.

Another possible risk that has been raised by scientists is recombination between a genome segment from a viral genome transcript and the genome of another challenge virus entering the plant genome. Recombinations occur frequently in nature, so the issue is whether there is something unique about the recombination of an artificial gene construct. Once again, Fuchs and Gonsalves write that "So far, no recombination event has been found in CP gene-expressing transgenic plants in the field."[15]

Horizontal gene flow from cultivated crops through pollen or other methods of gene transfer (transgene introgression) to compatible wild relatives was also raised as a potential risk of transgenic pathogen-derived resistance. C. Neal Stewart, Matthew D. Halfhill, and Suzanne I. Warwick have reported that "Transgenes engineered into annual crops could be unintentionally introduced into the genomes of their free-living wild relatives. The fear is that these transgenes might persist in the environment and have negative ecological consequences. Are some crops or transgenic traits of more concern than others?"[16] The authors recommend that large-scale genetic modification should be avoided in high-risk crops in which evidence of introgression has occurred and that the risks and benefits of transgene introgression should be studied on a case-by-case basis.

Finally, the food safety of virus-resistant plants has been discussed regarding allergenicity. When a virus-derived transgene is inserted to create a GMO, it can have amino acid sequences that can cause new allergens or cause enhancement of intrinsic allergens. Although this is a hypothetical risk, no adverse allergenic effects have been reported for papaya transgenic disease-resistant crops with a mild strain of PRSV. Nonetheless, some scientists express caution: "it is prudent to investigate food safety aspects of virus-resistant transgenic plants."[17]

From all the reviews on virus-resistant plants, there is no reliable evidence of any adverse human health effects from the consumption of such transgenic plants. One 2010 study on the compositional differences between transgenic and nontransgenic papaya for nutrients of beta-carotene and vitamin C as well as for two natural toxicants (benzyl isothiocyanate and carpaine) showed no significant differences.[18] Zhe Jiao, Jianchao Deng, Li Gongke, Zhuomin Zhang, and Zongwei Cai found that the compositional variability among papayas harvested across different time periods was higher than the compositional variability between transgenic and nontransgenic varieties. Other studies recommend that precaution and continuous oversight both pre- and postrelease should be maintained in the advent of unintended consequences.

With respect to environmental effects, another paper has reported that transgenic papaya had adverse effects on soil microorganisms. Xiang-dong Wei and colleagues studied soil properties, microbial communities, and enzyme activities in soil planted with transgenic and nontransgenic papaya under field conditions.[19] The transgenic papaya had a transgene consisting of a mutant gene of the papaya ringspot virus (PRSV), a neomycin antibiotic marker gene that confers antibiotic resistance to the genetically modified papaya cells, and a cauliflower mosaic virus promoter. The researchers collected soil samples before planting and after harvesting, undertook chemical and enzyme analysis, and enumerated colony-forming bacteria, actinomycetes, and fungi. They found that "transgenic papaya altered the chemical properties, enzyme activities, and microbial communities in soil."[20] Notwithstanding those effects, they observed that the transgenic papaya showed higher resistance to PRSV, had better growth, and produced more and larger fruits than the parental nontransgenic strain. Finally, the authors noted that "all these [observed effects] suggest the potential risk of field released GM plants, especially antibiotic genes like NPTII [neomycin phosphotransferase II marker genes] in GM plants, causing undesirable and unpredictable ecological effects."[21] The researchers conclude that genetically modified plants that are grown in the same soil for more than three months "could change the rhizospheric microbial metabolism; cause negative effects on soil quality, structure, and function; and affect enzyme synthesis and activity, as well as soil processes such as decomposition and mineralization of litter."[22]

The use of transgenes from pathogens to stimulate the plant's intrinsic immune system (pathogen-derived resistance) to protect itself from an invading pathogen has been approved in the United States for papayas, squash, and plums. Scientists combined the coat protein genes from three different viruses and created a squash hybrid with multiviral resistance. Similarly, genes from the plum pox virus were used to create disease resistance for the plum. Other transgenic crops in various stages of development for disease resistance include rice, wheat, apples, tomatoes, bananas, potatoes, barley, and soybeans.

Even with the widely reputed success of transgenic papaya, utilizing "pathogen-derived-disease-resistance,"[23] scientists have been developing nontransgenic methods using RNA silencing. According to one report, "Compared with the transgenic method for antiviral resistance, this approach is simpler, safer, environmentally friendly, and relatively inexpensive."[24] The authors used direct mechanical inoculation of papaya plants with bacterially expressed RNA that targets the gene of the papaya ringspot virus and interferes with the virus infection.

Chapter 9 discusses whether the potential for unintended consequences for transgenic crops is any greater than it is for traditionally bred crops.

7 Insect-Resistant Crops

Modern developments in biotechnology and transgenics brought new thinking to the methods of insect control and crop protection. In a way, the whole field of agricultural insecticides was turned inside out. The tradition, dating back centuries, had been to spray plants with chemical compounds that would kill or ward off insects that damaged crops. The revolution in thinking introduced the idea of building insecticides into the chemical structure of the plant.

After scientists understood the toxins that killed insects and the fact that toxins came from a biological source, then the possibility existed that genes coding for the toxic proteins could be spliced into the genome of the host plants. One of the potential benefits to this approach was that when the insecticide is built into the seed, farmers need to purchase only the seed and not make a separate purchase of chemical insecticides.

Another benefit is that plant-embedded insecticides are available at every stage in the crop's development and in every part of the crop, including the roots, stems, leaves, and fruit. Wherever the insects bite, that is where the insecticide is activated. A third benefit is that there are no chemical runoffs and occupational exposures that typically accompany applied chemicals. Field workers who apply chemical insecticides and are exposed to heavy applications are spared from a variety of illnesses.

Insecticides are reported to have been first used in agriculture about 4,500 years ago in Sumeria, where farmers used sulfur compounds to ward off insects and mites. Around 3,200 years ago, the Chinese applied mercury and arsenical compounds to their bodies to destroy body lice. In the millennia since, people have sought to repel or kill insects by using many other substances, such as smoke, tar, and pyrethrum from certain plants.

By the nineteenth century, farmers fought insect blights with a few inorganic chemicals. Paris green consisted of copper and lead compounds. The mixture, which originally was used against rats in the Parisian sewers, proved effective against crop insects. Another popular inorganic insecticide was Bordeaux mixture, which was comprised of copper sulfate, lime, and water. Farmers also used arsenic and kerosene emulsions for crop protection.

It has been estimated that about 67,000 pest species are capable of damaging crops and that about 9,000 of those species consist of insects and mites. Globally, crop losses are at around 37 percent of what is planted, and about 13 percent of those losses is attributed to insects.[1]

By the early twentieth century, farmers used a variety of organic compounds—such as phenols, creosote, and petroleum products—to address insect and fungus blight. Modern synthetic organic insecticides were developed after World War II, and in the 1950s and 1960s, compounds such as aldrin, chlordane, DDT, dieldrin, endrin, and parathion became widely used until some were banned or highly restricted because of their toxic effects.

As previously noted, in the aftermath of the recombinant DNA revolution, it became possible to extract a gene coding for a biological toxin from one species and incorporate it into another. The two most promising classes of toxins or insecticidal proteins for delivery into plant cells were lectins (insecticidal plant proteins) and bacterial toxins. In order to integrate a biological insecticidal toxin into the germplasm of a plant, it must be a protein. Copper or arsenic cannot be incorporated into the germplasm because there is no gene for these substances. Some of the first bacterial protein toxins to be considered for creating a transgenic insecticidal plant were from the soil bacterium *Bacillus thuringiensis*.

Bacillus thuringiensis (Bt) was discovered by the Japanese biologist Ishiwata Shigetane in 1901. Shigetane isolated the bacterium from dead silkworm larvae when he was investigating the cause of "sotto (sudden collapse) disease," which was highly destructive to the silk industry. He believed that the bacterium was responsible for the death of silkworms in Japan and named it *Bacillus sotto*. He did not recognize Bt as a pesticide.

A decade later, in 1911, a German scientist, Ernst Berliner, rediscovered a related strain of Bt from dead Mediterranean flour moth larvae found in a flour mill. He reported the existence of a crystalline protein in the bacterium, which he believed caused the disease in the caterpillar. He named the bacterium *Bacillus thuringiensis* because it was found in the German state of Thuringia. The mode of action of Bt was not discovered until much later.

In 1976, Robert Zakharyan and his colleagues reported the presence of a plasmid (a circular piece of DNA) in a strain of Bt and hypothesized that it played a role in the creation of an endospore (a dormant, tough, nonreproductive structure produced by certain bacteria) and the formation of a crystalline protein.[2] In the next decade, Bt was studied extensively, and scientists learned that more than 200 types of Bt proteins exhibit some degree of toxicity to specific insects.

The mode of action of Bt's insecticidal properties is reasonably, albeit not completely, understood. We begin with the fact that Bt has two phases to its life cycle. It can undergo *cell division* or *binary fission*, or it can undergo *spore development* or *sporulation*, which is the formation of the heat-resistant endospores located at one end of the rod-shaped bacterium. Spore formation is a method that the bacterium uses in order to protect its reproductive cycle for more favorable environmental circumstances. Starvation of the bacterium initiates sporulation. Under the right conditions, the endospore can be activated to revive the bacterium, a process called *germination*.

As part of Bt sporulation, there is a formation of certain crystals (*parasporal crystals* are crystalline protein that forms around a spore). The proteins in these crystals are referred to as *Cry proteins* (from crystal endotoxins) and usually are located on a plasmid in the bacterium. The Cry proteins exhibit insecticidal properties to moths, butterflies, flies, mosquitoes, beetles, wasps, bees, ants, sawflies, and nematodes. Each of the Cry insecticidal proteins is highly specific to different insects.

When susceptible insects or their larvae ingest the crystals, their alkaline digestive tract turns the insoluble crystal to a soluble form. The enzymes in the insect's alkali gut free the protein toxins from the crystal, whereupon they create pores in the cell membranes of the insect, lysing (breakdown of the cell membrane) midgut epithelial cells. When the midgut epithelium releases the lysed cells' contents, it provides a germinating medium for live bacteria, leading to the disruption of membrane integrity and resulting in the death of the insect.[3]

Farmers began using Bt in the 1920s. By 1928, a spore-based formulation of Bt under the trade name Sporine was commercialized in France. Twenty years later, Bt was available for farmers in the United States. Bt is one of the few pesticides that have been approved (both as granules and as a liquid) under the organic standards as a *microbial pest control agent* (MPCA).

Among the several major crops that were genetically modified in the laboratory with Bt δ-endotoxins (the pore-forming toxins produced by Bt bacteria) and were approved for field testing are alfalfa, corn, cotton, cranberry, eggplant, poplar, potatoes, rice, spruce, tobacco, tomatoes, and walnuts. EPA approved the first commercial release of a transgenic insecticidal crop in May 1995 with potatoes, corn, and cotton.[4] By that year, there were 182 Bt products registered by EPA.[5] Bt potatoes were introduced by Monsanto to control the Colorado potato beetle. By 2003, Bt canola, papaya, soybean, and squash had entered commercial markets. EPA and FDA reached a consensus that the Bt δ-endotoxins expressed in the crop do not represent a human health hazard.

The specificity of Bt toxins for explicit insects makes them highly desirable biological control agents. Instead of spraying Bt over the crops, where it can be inactivated by sunlight or washed off, Bt transgenic crops use only one of two of the Bt toxins. The toxicity of Bt is receptor-mediated. This means that for any organism to be affected by the Bt toxin the way it is designed to be used against insects, it must have specific receptor sites in the insect's gut to which the protein can bind. It is generally understood that humans and the majority of beneficial insects do not have the Cry toxin receptor sites. According to Anthony M. Shelton, Jian-Zhou Shao, and Richard T. Roush, entomologists applauded insect-resistant technology: "Transgenic plants expressing insecticidal proteins from the bacterium *Bacillus thuringiensis* (Bt) are revolutionizing agriculture."[6]

Although humans and other animals may not have the receptor sites of Lepidoptera, that does not preclude other means through which Cry toxins may affect them. Any single protein may interact in many pathways—not just one. For example, a Cry protein could potentially induce allergenicity in mammals but not have this effect in insects. Kun Xue, Jing Yang, Biao Liu, and Dayuan Zue note: "In current commercial GMOs, insertion of exogenous DNA sequences into the plant genome is random....the transgenes, their products, or the changed biochemical pathways may also interact with other genes or pathways....it remains very challenging for regulators to get information on unintended changes outside of a narrow range of agronomically relevant metabolites that are profiled."[7]

The StarLink Case

In 1987, Plant Genetics Systems (PGS) became the first company to produce genetically engineered crops for pest protection. The company's research team, led by Mark Vaeck, introduced transgenes from the soil bacterium *Bacillus thuringiensis* (Bt) in tobacco plants. The genes encode an insecticidal protein toxin that protects tobacco crops from the tobacco hornworm larvae. This was the first successful attempt to modify tobacco plants genetically with a Bt toxin gene against the larvae of the tobacco hornworm.[8] PGS was a joint venture between Advanced Genetic Systems (AGS) of Oakland, California, and European investors. One of the companies that were interested in this new technology of insect-resistant plants was Hoechst Schering AgrEvo GmbH (AgrEvo). AgrEvo was a German manufacturer of chemicals for crop protection, which included herbicides, insecticides, and fungicides. In 1995, AgrEvo decided to shift from crop protection to crop production and acquired Plant Genetic Systems, which had developed a transgenic variety of corn utilizing one of the toxic proteins in *Bacillus thuringiensis* called Cry9c.

The genetically modified corn, with the trade name StarLink, consisted of two transgenes—one transgene that resists the herbicide glufosinate and one transgene that encodes the Bt protein Cry9c, which is toxic to the European corn borer. A five-year period of regulatory reviews and approvals started with PGS, continued with AgrEvo, and finally ended with Aventis, which became the successor of AgrEvo (box 7.1). The three U.S. agencies that were involved with the oversight of StarLink corn for field tests and commercial products were the Food and Drug Administration (FDA), the Environmental Protection Agency (EPA), and the Animal and Plant Health Inspection Service of the U.S. Department of Agriculture (APHIS/USDA).[9]

Plant Genetic Systems received approval from APHIS for field trials of StarLink corn covering 1996 to 2000 under its notification and acknowledgment procedure. The EPA had regulatory oversight over genetically modified organisms with insecticidal proteins by its authority under the Federal Insecticide, Fungicide, and Rodenticide Act of 1996 (FIFRA). GMO crops with transgenic insecticidal proteins were classified as *plant-incorporated protectants* (PIPs), a new variety of pesticides.

Under the 1996 Food Quality Protection Act, the EPA could not approve a pesticide unless it had a "reasonable certainty the pesticide will not cause

Box 7.1

The Regulatory History of StarLink Corn, 1996 to 2001

1996: The Animal and Plant Health Inspection Service of the U.S. Department of Agriculture (APHIS/USDA) approves field trials for StarLink.

March 1997: The Environmental Protection Agency (EPA) issues an experimental use permit to field-test StarLink in twenty-eight states. After the trials, all crops would be destroyed or used for research.

September 1997: EPA announces that Plant Genetics Systems has asked to register StarLink as a pesticide that is not limited to nonfood uses.

February 1998: Plant Genetics Systems petitions APHIS to deregulate StarLink.

April 1998: Plant Genetics Systems amends its petition to EPA (for an exemption only in corn that is used for feed) to include corn that is used for feed as well as corn that is used in meat, poultry, milk, or eggs resulting from animals fed StarLink corn. EPA issues a temporary exemption of tolerance requirements for StarLink residues in feed and food products that are derived from animals that have eaten the food.

May 1998: EPA approves the registration of StarLink as limited to nonfood uses. EPA grants an exemption of a tolerance requirement for pesticide residues of StarLink in feed and food products that are derived from animals that ate the food. APHIS issues nonregulated status to StarLink. AgrEvo completes its consultation with the Food and Drug Administration (FDA). StarLink in animal feed does not raise issues of premarket review or approval by the FDA.

April 1999: EPA announces that AgrEvo has asked for an expansion of the tolerance exemption to cover direct human consumption of food with Cry9c in addition to feed.

February 2000: The EPA's Scientific Advisory Panel on the Federal Insecticide, Fungicide, and Rodenticide Act of 1996 (FIFRA) meets and concludes that there is "no evidence to indicate that Cry9c is or is not a potential food allergen."

July 2001: EPA rejects Aventis's petition to allow traces of StarLink to remain in the food supply.

Source: Donald L. Uchtmann, "StarLink: A Case Study of Agricultural Biotechnology Regulation," *Drake Journal of Agricultural Law* 7 (2002): 159–211.

harm." Under this standard, a manufacturer had to show that the genetically modified crop (in this case, a PIP) was not toxic to consumers and was not allergenic.

Plant Genetic Systems divided its application into two permit requests—one for animal feed and another for food. In its data to the EPA, the company reported that its transgenic corn variety, StarLink, with the toxic protein Cry9C, was tested for allergenicity by putting it into an acid solution that simulated the human gut and subjecting it to heat. Testing StarLink directly on human subjects was not considered ethical, especially before it passed the gut-simulated conditions. Compared to other Bt proteins, Cry9C persisted longer. It is generally recognized in the field of allergenicity that known allergens persist in the digestive tract longer than is typical for most proteins. Because the company did not carry out more detailed studies that showed StarLink corn was not allergenic to humans, the variety was given limited approval by EPA in 1998 exclusively for animal feed.

The EPA instructed the company to warn farmers and distributors that they should not commingle different Bt corn varieties and that farmers must grow the variety on an isolated field so that the corn pollen would not fertilize other corn varieties. Through a merger in 1999, AgrEvo became Aventis Crop Science. Its parent company Aventis SA submitted data to the EPA that StarLink was not allergenic and petitioned for an exemption that would allow StarLink to be used directly in human food. EPA's Scientific Advisory Board (SAB) met in February 2000 and did not grant Aventis an exemption for Cry9c. The SAB stated that "there is no evidence to indicate that Cry9c is or is not a potential food allergen."

Meanwhile, nongovernmental organizations in the United States were petitioning the EPA to require labeling of Bt products because they were skeptical about the sequestration of corn cultivars. Someone working for Friends of the Earth purchased corn products at a market in Silver Springs, Maryland, and the group had the samples tested by Genetics ID, an Iowa-based laboratory, for signs of Cry9c. The tests revealed traces of Cry9c DNA in one of the products of taco shells, which brought international media attention to the failure of sequestration of the Bt corn products. In October 2000, the FDA confirmed that StarLink corn was in the food chain in consumer corn products. By November 2000, companies began recalling corn products after the FDA mandated the recall.

Some have argued that Cry9c proteins might not have been in the corn products because the testing was done for DNA and proteins decay more easily in processing. After a StarLink recall and national publicity about the taco shells, fifty-one people reported to FDA that they had experienced adverse effects after consuming the taco shells. The Centers for Disease Control and Prevention (CDC) undertook a study and found that twenty-eight reports could possibly have been related to the StarLink proteins. But after studying the blood of the twenty-eight individuals, the CDC concluded there was no evidence that their reactions were associated with allergenicity to Star-Link Bt Cry9c protein. The EPA's Scientific Advisory Panel reviewed the CDC results and stated that "the technical approach for the detection of Cry9c protein and antigen specific IgE's [antibodies triggering an allergic reaction] is limited and cannot resolve the issue of the presence or absence of Cry9c-specific IgE in the serum of individuals reporting adverse reactions after eating corn."[10] Notwithstanding the uncertainty over the allergenic effects, the positive findings of Cry9c DNA in corn products revealed that the EPA requirements for StarLink corn were not followed, fueling public mistrust in regulatory oversight of GMOs.

Bt has been studied as an insecticidal spray for its human toxicological effects with overall favorable results. When Bt toxins were genetically embedded in plant germ plasm, there were other risk considerations. The transgenic Bt plant contains one or more Cry toxins through the entire plant. Among the questions raised were those concerning the types of human exposures involved in consuming parts of the plant, other possible protein changes in the plant resulting from the transgene, and the ways that the Cry proteins function in the human gut.

Several systematic reviews have been written evaluating the weight of evidence on the toxicology of Bt transgenic plants. A systematic review prepared by scientists of Monsanto concludes with the following comment:

> Cry Bt proteins, whether in microbial pesticide products or expressed in Bt crops, have been used and consumed safely for decades. The levels of Cry Bt protein in GM crops are very low and are often reduced further by food processing. In addition, extensive testing of Bt proteins, single-Bt trait crops, and stacked trait crops containing Bt proteins has not revealed any harm to non-target insects and other non-target species, including humans.[11]

Another paper published scientists at Dow Agro Sciences states that, because of European regulations, "proteins developed for insect resistant crops undergo extensive testing for mammalian safety" and follow "a tiered approach that examines familiarity and relatedness to other proteins, mode of action, and direct testing for mammalian toxicology." They conclude that animal studies of Bt corn have shown no evidence of acute toxicity.[12]

Nestor Rubio-Infante and Leticia Mereno-Fierros cover an extensive body of research without preselection based on outcome and state that "The majority of the laboratory studies that were performed to test the infectivity and toxicity of Bt commercial products have indicated that these products are safe." Yet they conclude that "such studies are not enough proof that these products are innocuous to mammalian cells or vertebrate organisms."[13] The reasons behind their skepticism are that long-term studies must be completed; that the effects of the proteins on the gastrointestinal tract, the immune system, or the genitourinary tract have yet to be determined; and that subchronic and chronic studies in humans should be performed. Other researchers cite a three-generation feeding study in rats with Bt insecticidal maize that reports that "some minimal histopathological changes in liver and kidney"[14] in the GM-fed animals should not be ignored. Michael N. Antoniou and Claire J. Robinson write that "examination of the detail of the study reveals that the GM-fed rats suffered damage to their liver and kidneys and alterations in blood chemistry, which some scientists may view as unresolved safety questions demanding further inquiry."[15]

Xue and colleagues used a proteomics approach to study Bt rice compared to its non-GMO counterpart. They report that "transgenic rice lines were found to differ in their protein contents from their non-GM counterparts, which could raise concerns regarding their potential risks for human health and ecology."[16]

In conclusion, Bt crops have become ubiquitous in the United States and other countries such as China and India, and except for the StarLink affair, there has been no other singular event that threatens their future. Studies report that Bt crops have shown high profitability, increased production, and reduced insecticide applications, up until the build-up of insect resistance to Bt.[17] The rise of insect resistance to Bt is being addressed by the pyramiding of Cry toxins, the development and implementation of refugia for nontransgenic counterparts within or around a Bt field, and the use of traditional management techniques in conjunction with transgenic crops. For some reviewers, there remain uncertainties about the chronic and long-term effects Bt-crop Cry proteins, the nontarget effects of Bt for certain species, changes in secondary insect pests, and the impacts of Bt crops on honey bee navigation. The overuse of BT is likely to undercut its selective use in organic farms.[18] Other scientific groups back the human safety of Bt proteins with a "high degree of certainty."[19]

The next chapter examines how an understanding of genetic mechanisms helps to frame the risk assessment of genetically engineered crops and explain some of the disagreements on GMO safety among scientists.

8   Genetic Mechanisms and GMO Risk Assessment

This chapter focuses on the reasons that scientists differ on the risk assessment of GMOs by examining their presuppositions about the mechanisms involved in genetic technology. American and European approaches to the risk assessment of GMOs differ because regulatory authorities do not share a common framework for and expectations about the uncertain effects of molecular breeding compared to traditional breeding.

The history of science explores the evolution of theories and models from their earliest to their mature stages of development. For example, the model of the atom proposed by Niels Bohr during the nascent stages of quantum mechanics was viewed as similar to a solar system, with a nucleus (protons and neutrons) and electrons orbiting in quantized orbits. It accounted for the emission spectra of the hydrogen atom but had to be refined to explain the spectra of other atoms. The theories and models in certain subfields of science are the basis for any risk assessment that might be informed by those subfields.

The models of plant and human genomes, like the Bohr atom, have evolved from earlier simplistic and idealized conceptions to more nuanced and realistic structures, which were grounded in the empirical findings of legions of molecular geneticists. Initially, the model of the genome was static—like a Lego structure. A gene comprised of DNA was a fixed entity that encoded RNA, which cascaded into a protein. It was the central dogma of molecular genetics: DNA → RNA → Protein. Soon thereafter, scientists found that they could reverse the order and move from RNA to DNA by the use of an enzyme. Genomes also have noncoding DNA sequences that are not involved in the synthesis of proteins. Initially, they were called *junk DNA*—the flotsam and jetsam of evolution that no longer served a useful function. This also proved to be simplistic because noncoding DNA was found to have a variety of functions.[1]

Early views about plant or animal genomes held that a foreign gene construct (gene, promoter, marker) that is inserted into the chromosome would simply replicate its function in the organism from which it was extracted or become a functionless appendage—junk DNA.

After the discovery of recombinant DNA, genes were considered universally fungible in that they could be moved around from one biological system to the next across different species, genera, or kingdoms without affecting the transplanted gene or the other genes in the genetically modified organism. As in many inchoate theories, this also proved to be far too simplistic. Moving a gene from one system to another requires a special promoter that tells the gene how to turn on. In addition, the current methods of moving the gene into a new species does not guarantee where it will be found in the chromosome. Its placement in the chromosome could affect the expression of the gene (called the *position effect*). To make things even more complex, the genome in which the foreign gene is placed acts more like an ecosystem than a Lego system. That means that genes interact with one another. If a gene is introduced into a plant genome, some other gene (or genes) already in the host plant could be affected. That preexisting gene might code for a protein that is a nutrient, a toxin, or an enzyme, or it might perform a regulatory function.

Even the concept of the gene as the fixed template for protein production has been questioned by authors like Evelyn Fox Keller, who, in her book *The Century of the Gene*, writes, "In higher organisms DNA sequence does not automatically translate into a sequence of amino acids, nor does it, by itself, suffice for telling us just which proteins will be produced in any given cell or at any stage of development."[2] In a review of Keller's contribution to an edited book, Evan Charney concludes that "the idea of a predetermined code of DNA as the rigid template of heredity in which our fates are transcribed is at odds with phenotypic plasticity, the ability of organisms to adapt to the demands of their particular ecological niches. This plasticity extends to the DNA molecule itself."[3]

The field of epigenetics introduces another layer of complexity in the genetic model. Proteins interwoven into the DNA may determine the conditions under which a gene can serve as the template for the synthesis of a protein product. Epigenetic marks function as switches that turn the gene function on or off.[4] And Barbara McClintock's discovery of jumping genes in maize suggests that the genome is not fixed and stationary but can be altered and rearranged.[5]

The ecosystem model is well understood by scientists who explain gene-gene interactive effects. H. Vaucheret and colleagues discuss the complexity of integrating a foreign gene into a host genome:[6] "transgenes can undergo silencing after integration in the genome. Host genes can also be silenced as a consequence of the presence of a homologous transgene.[7] ... The transfer of methylation and silencing from one locus [a specific location of a DNA sequence on a chromosome] to another clearly indicates that independent parts of the genome communicate and exchange information."[8]

When DNA complexes are transferred to plant cells through rDNA (ribosomal DNA) techniques via *Agrobacterium tumefaciens* or gene guns, they enter different locations on the chromosome. The protein expression can be affected by varied mechanistic events at different locations, such as alternative splicing of RNA (a single DNA sequence may code for multiple proteins) and chromatin structure affecting gene expression; sequences far from the coding sequence; and posttranslation events. Although the plant cells in culture can be selected for the transgene, given the possibilities during transcription, protein expression can vary.

Just as we must understand the mechanism of the atomic nucleus to appreciate the risks of bombarding atoms with elementary particles, so too it is critical that the mechanisms of the genome be understood when we seek to understand the effects of introducing foreign genes into a plant by unnatural processes. Differences in how some scientists view the risks of transgenic crops may be rooted in the assumptions that they hold about the plant genome.

I begin this chapter by first stating some generally recognized principles in plant genetics. Starting with simple concepts, I proceed to introduce more complexity. Then I explore some of the areas of contested knowledge about risk and identify the presuppositions about the opposing claims. Where there is uncertainty about the science behind the claims, I make that transparent.

David Schubert, a geneticist and professor at the Salk Institute, cites three important conditions underlying *transgenics* (the movement of genes across organisms) that I find useful for my analysis:[9]

- The introduction of the same gene into two different types of cells can produce two very different protein molecules.

- The introduction of any gene, whether from a different or the same species, usually can change overall gene expression and therefore the phenotype of the recipient cell.

- Enzymatic pathways that are introduced to synthesize small molecules, such as vitamins, could interact with endogenous pathways to produce novel molecules. This is particularly relevant to creating transgenic nutritional changes in plants.

These points are not contested in the scientific literature, but they highlight the ecosystem concept of the plant genome and cellular environment. What is contested is their significance for distinguishing traditional breeding from molecular breeding. One issue that has drawn a great deal of discussion is that both traditional breeding and molecular breeding involving chemical mutagenesis can give rise to hazardous products. Therefore, why should molecular breeding be singled out for special regulatory oversight? Why should the focus of risk assessment be on the method of breeding (molecular breeding) rather than the product of breeding (the modified embryo)?

A particular method of breeding might be rationally singled out if it is associated with a higher probability of unanticipated or hazardous results or if its results, whether more or less probable, yield more serious dangers than other methods. As previously discussed, this is referred to as the process versus product debate in biotechnology. Some scientists have criticized the idea of transgenic exceptionalism attributed to molecular breeding. Kent J. Bradford and his colleagues have written that "Long-accepted plant breeding methods for incorporating new diversity into crop varieties, experience from two decades of research on commercialization of transgenic crops, and expanding knowledge of plant genome structure and dynamics all indicate that if a gene or a trait is safe, the genetic engineering process itself presents little potential for unexpected consequences that would not be identified or eliminated in the variety development process before commercialization."[10]

There are two notable points in this paragraph. First, if you know the gene is safe in one system, it will be safe when you transport it to another. Second, if there are any untoward effects of the new transgenic plant, they will be eliminated by the breeders before commercialization because, it is argued, any unexpected consequences will be immediately visible or chemically detectable and easily dealt with. The first point is not consistent with Schubert's three principles because a transplanted gene can have unanticipated effects in the new organism. The second point needs to be discussed in the context of regulation. The plant breeder may have found that the transgenic plant meets all the agronomic goals. It is not obvious that some negative changes to the plant in the form of nutritional pathways will always be picked up by the breeder. Confidence that the breeder will pick up any potential adverse effects of the new transgenic crops is resolved not by science but only by trust in breeders.

This idea of breeder oversight is emphasized in Nina Federoff and Nancy Marie Brown's *Mendel in the Kitchen: A Scientist's View of Genetically Modified Foods*: "No matter which promoter is used, precisely where the DNA is inserted into a plant's genome is more or less random. A few insertions affect the plant's ability to grow and be productive. These plants and their progenies are discarded as well. For this reason, breeders create many transformed lines. ... The best plants in the field trials are saved; the rest are discarded."[11]

In its 2004 report on the safety of genetically modified foods, the National Academy of Sciences (NAS) (now the National Academies of Sciences, Engineering, and Medicine) made some relevant observations about the issue of the relative likelihood of compositional changes in plants comparing conventional and molecular breeding methods. According to the NAS report, the likelihood of changes among breeding methods falls into a continuum: "All evidence evaluated to date indicates the unexpected and compositional changes arise with all forms of genetic modification [such as pollen-based crossing], including genetic engineering."[12] The NAS report stated that there is not a simple dichotomy of breeding methods, traditional versus molecular breeding, in assessing unanticipated effects and hazards. Twelve examples are given of breeding methods, some involving rDNA (transgenic) and some involving radiation and chemicals to induce mutagenesis. The NAS considers the methods that are most likely to produce unintended health effects to be those using chemical and radiation to induce mutagenesis. George Acquaah, in his textbook titled *Principles of Plant Genetics and Breeding*, writes, "Invariably, the mutagen kills some cells outright while surviving plants display a wide range of deformities. Even those plants with the desirable mutations always inherit some undesirable side effects."[13]

The risks of unintended effects using rDNA, according to the NAS, can be rated low if the foreign genes come from closely related species and the vector is *A. tumefaciens* (fourth lowest) or much higher if the rDNA is used to transfer genes from distantly related species (next to the highest). The NAS does acknowledge that mixing genes from distantly related species increases the chances that there will be unintended health effects. The assessment of relative risk of the likelihood of unintended health effects is not a precise science. A group of plant geneticists create a taxonomy of risks based on their accumulated knowledge and examples that come to mind. There is no database of breeding methods and their unintended consequences.

The National Academies of Sciences, Engineering, and Medicine (NASEM) is a highly respected science body that draws on experts in relevant fields when it assembles a panel study. Also, the draft of the study is carefully reviewed by panel members of the National Academies and by scientists outside the designated committee who are assigned to the report. A NASEM report is supposed to represent the consensus viewpoint of the scientific community on an issue. The conclusion of the NASEM is that all forms of genetic modification (including traditional breeding and molecular breeding) "may potentially lead to unintended changes in composition, some of which may have adverse health effects."[14] And NASEM continues to emphasize that molecular breeding using rDNA techniques is not inherently hazardous.

There is yet another perspective that believes unintended effects will occur more frequently from natural genetic exchanges: "The reality is that 'unintentional consequences' are much more likely to occur in nature than in biotechnology because nature relies on the unintentional consequences of blind random genetic mutation and rearrangement to produce adaptive phenotypic results, whereas GM technology employs precise, specific, and rationally designed genetic modification toward a specific engineering goal."[15] This statement, which was signed by fifteen scientists from high-ranking universities, seems to be at odds with the NASEM position, which implies that genetic exchanges from organisms that interact in nature will likely produce fewer unintended health effects than those that are introduced by human mechanisms, whether chemical mutagenesis or rDNA, especially with taxonomic proximity between recipient and host plants.

Where there is disagreement among scientists, it is over the potential risks from molecular breeding that transfers foreign genes from distantly related organisms, as well as over the requirement for special promoter sequences and *marker genes* (antibiotic-resistant genes) that raise the risks of untoward effects. There is no definitive science to evaluate the relative risks. There are only the judgments of seasoned plant geneticists, chemists, breeders, and their critics on the relative importance of random mutagenesis or traditional breeding versus the allegedly more precise transgenics from distantly related species. Jonathan R. Latham, Allison K. Wilson, and Ricarda A. Steinbrecher argue that transgenic plants are sometimes going to experience mutations with phenotypic effects: "Both insertion-site and genome-wide mutations may result in transgenic plants with unexpected traits…these unexpected traits range from altered nutrient or toxin levels to lower yields under certain environmental conditions."[16]

Precision and Safety

One popular contention about the benefits of molecular breeding over traditional breeding was based on the precision and accuracy of the former over the latter. For one thing, in molecular breeding, the gene inserts are known, whereas in traditional breeding, there are bundles of genetic exchanges that are not fully accounted for. Robert May, an Australian scientist who was chief scientific adviser to the UK government, president of the Royal Society, and professor at Sydney, Princeton, Oxford, and Imperial College London, stated during a BBC interview that "GM techniques which in the precise and targeted way bring in a couple of genes that you know what they do and you know where they are is vastly safer, vastly, vastly more controlled than the so-called conventional breeding."[17] Other scientific studies highlight the uncertainties associated with the methods of molecular cloning. Hae-Woon Choi, Peggy Lemaux, and Myeong-Je Cho report a higher degree of *aneuploidy* (an abnormal number of chromosomes in a cell by an extra or missing chromosome) in producing transgenic oat in contrast to nontransgenic oat.[18]

If a theory does not provide a clear answer to the safety or risks of GMOs, perhaps practice and experience can. There is no disagreement that molecular breeding can produce a harmful product. The questions are, How frequent, how harmful (or at what level of harm), and at what probability? Some traditionally bred potatoes have high levels of glycolalkaloids, which can make consumers sick. Federoff and Brown argue that experience reveals that very few gene insertions have a phenotypic effect on plants: "Indeed, plant geneticists have been surprised at how few gene insertions have a visible effect on plants. Scientists studying the little laboratory weed (or a model plant in plant biology for its relative small genome size and short life span) *Arabidopsis thaliana* have used *Agrobacterium* to make thousands upon thousands of T-DNA (Transfer DNA) insertions in or near genes to see what effect disrupting a particular sequence has on the plant. Most of the time, inserting a new gene has no effect at all; the plants grow normally. A few insertions cause mutations, but for every hundred only one or two make a visible change in how the plant looks, grows, or reproduces."[19]

For consumers of plants, it is not so much how the plant looks but what biochemical changes have occurred in the plant because of the transgene cassette. Point two in Schubert's list suggests that the foreign gene insert results in changes in gene expression. It would be valuable to have data on how frequently a gene insert changes the nutritional components or microtoxins in a plant. Based on the frequency of such changes, what type of monitoring is necessary to ensure consumer confidence? Some scientists place their confidence in "what the gene is and what it codes for"[20] rather than on what gene-gene interactions, gene-cellular environment interactions, or existing enzymatic pathways could modify or what new enzymatic pathways might emerge—in other words, on the unintended outcomes.

Approaches to Risk Assessment

Some form of risk assessment has been a part of introducing GMOs into commerce. The fundamental principles underlying such as assessment are generally shared, but the methods vary significantly. As noted by Claudia Paolotti and her colleagues: "The essence of the safety approach is that the new [biotechnology-derived] food (or component thereof) should be compared with an appropriate conventional counterpart, that is with a food already accepted as safe based on its history of safe use as food."[21] Under the risk assessment guidelines adopted by the Codex Alimentarius Commission, established by the United Nations Food and Agricultural Organization and the World Health Organization, if all significant differences between the GMO and its conventional counterpart can be identified and found safe for human health, then the GMO can be considered to be as safe as its conventional counterpart and declared a "substantial equivalent" to the non-GMO counterpart.

The European Food and Safety Authority (EFSA) was established to provide an independent scientific risk assessment on the health and safety of GMOs by interpreting the meaning of the general goals and issuing recommendations. In the European context, "substantial equivalence" is the outcome arrived at after a study is made on the GMO. According to a Monsanto scientist, "Substantial equivalence is a concept that provides guidance by helping to identify the questions to be asked during the safety assessment of new foods, feeds or processed fractions; it is not an endpoint."[22] EFSA evaluates each GMO by acquiring data on the composition, toxicity, allergenicity, and nutritional value. A selection is made on selected proteins in the GMO for study and comparison. Even though the GMO is not "chemically identical" to its non-GMO counterpart, it can, according to EFSA, for the purpose of human health and safety, still be declared "substantially equivalent." Other than compositional analysis, EFSA also recommends the use of animal feeding studies in certain circumstances to detect toxicologically or nutritionally relevant differences.

The U.S. approach to the risk assessment of GMOs advances goals that are similar to the goals of the European Union (EU) but takes a markedly different approach. The U.S. Food and Drug Administration leaves to the food manufacturers (the molecular breeders) the responsibility of doing the risk assessment for GMOs. The agency offers the food developers a "voluntary consultation" process, which means they may choose to consult with FDA scientists regarding tests on the composition (toxicology, allergenicity, and nutritional levels) of the GMOs. There are no published standardized tests required in this consultation process. Among its recommendations to GMO food developers, the FDA lists the acquisition of data comparing the composition of the bioengineered food to that of its parental non-GMO source.

The starting point for risk assessment differs significantly between the United States and the European Union. The FDA assumes that foods developed by the addition of foreign genes are *generally regarded as safe* (GRAS) (that is, substantially equivalent) unless proven otherwise, whereas in Europe, the designation of GRAS has to be demonstrated after testing is complete. If data on the compositional analysis of a GMO fail to demonstrate safety or substantial equivalence, then animal testing is suggested—first a ninety-day study and then long-term studies (if needed) for evaluating chronic effects.

Animal studies for whole foods have well-identified limitations for measuring human health risks.[23] They may yield false positives because of dietary imbalances in the test animals. Andrew Bartholomaeus, Wayne Parrott, Genevieve Bondy, and Kate Walker summarize the problems: "It is recognized that animal toxicology studies intended for human risk assessment have inherent limitations and challenges due to the use of a surrogate species and relatively small numbers of animals in each study group. Differences in physiology (toxicokinetics, toxicodynamics, nutritional requirements, and other biochemical characteristics), anatomical structure, and behavior can lead to positive or negative findings that may not accurately reflect human responses."[24] Also "increasing the duration of a WF [whole-food] toxicity study from 28 or 90 days to a long-term chronic study cannot be expected to correct for the inherent limitations of WF studies."[25] Their conclusion is that "WF animal toxicity studies are unnecessary and scientifically unjustifiable." These authors believe that, unlike methods used in animal studies, the most reliable methods for detecting unsafe GMOs is through the compositional and agronomic assessment of new GM crops. But they leave unanswered important questions: Who undertakes the compositional analysis? How many proteins, metabolites, RNAs, and DNAs are evaluated in the GMOs and compared with the parental strain? These questions are addressed in chapter 11, which discusses omics analysis.

Some scientific groups believe that whole-food animal feeding tests will provide reliable information about the risks of GMOs. Other groups consider that such tests will provide unreliable information about the risks of GMOs and believe that a comprehensive compositional analysis, including proteins and metabolites, is necessary for discovering any risks and unintended outcomes. It is now generally understood from the science that metabolites and proteins from GMOs can differ from their non-GMO counterparts. Conventional breeding can also result in plants with different metabolites and proteins. Does molecular breeding result in changes that are qualitatively distinct from the changes of traditional breeding? And if so, do those changes increase the risks of untoward properties in the food?

Some analysts accept only the weight of evidence rather than any individual provocative study showing GMO risks: "We argue that the totality of the evidence should be taken into account when drawing conclusions on GMO safety instead of farfetched evidence from single studies with a high risk of bias."[26]

This chapter has argued that the growing complexity of the plant genome must be accounted for in the risk assessment of GMOs. Simplified models of the genome understate the possibility of unanticipated events from transplanting foreign genes. Increasingly, scientists are more favorable to a comprehensive analysis of the molecules and metabolites in GMOs in comparison to the non-GMO counterparts. Heiko Rischer and Kirsi-Marja Oksman-Caldentey argue that omics analysis offers the best chances for detecting unintended effects in GMOs: "One of the major challenges is how to analyze the overall metabolite composition of GM plants in comparison to conventional cultures, and one possible solution is offered by metabolomics. The ultimate aim of metabolomics is the identification and quantification of all small molecules in an organism; however, a single method enabling complete metabolome analysis does not exist."[27]

No regulatory agency currently requires such an analysis, and there are no standards for undertaking it. For example, as of 2013, the European Union requires a mandatory ninety-day feeding study in rodents on whole GM food or feed in order to identify adverse effects. U.S. regulators do not have required risk studies but leave it to the food producers.

9 Contested Viewpoints on the Health Effects of GMOs

Thus far, I have explored the mainstream scientific literature on the principles of traditional and molecular breeding, pointing out where there are contested viewpoints. I have discussed how the methods of breeding may introduce unintended effects and questioned whether molecular breeding introduces any unique health or environmental impacts from the crops produced and foods derived from them. Looking into competing scientific interpretations may help to explain why there are wide divisions in society over GMOs. It is too easy to say that one group follows the science and the other group follows an ideology. That leads some observers to embrace the idea of "GMO deniers," referring to people who leave the science behind in favor of an irrational (or groundless) opposition to genetically modified food. But there is a scientific record of studies that support honest skepticism. Also, European and American scientists see the issues and the risks differently, which can explain why their respective regulatory systems are distinct.

In this chapter, I explore contested principles held by scientists that reflect not so much different interpretations of the same science but rather alternative presuppositions about how to apply the knowledge of plant genetics to the questions of risk and why scientists may reach different conclusions. There is a great deal that plant biologists agree upon, which is covered in the previous chapters. Here I focus on principles or conclusions for which there is lack of consensus—at least among certain groups of scientists. By focusing on the contested interpretations of general principles, I will be able to highlight the locus of disagreement.

Contested Point 1: Traditional and Molecular Breeding *Are* (*Are Not*) Qualitatively Distinct.

The mainstream position—based on reports of the National Academies of Sciences, Engineering, and Medicine (NASEM), government agencies, and professional agricultural organizations in plant breeding—is that different forms of breeding may produce unpredictable and unexpected effects and that there is nothing inherently riskier about molecular breeding using genetic engineering than breeding involving wide crosses, embryo rescue, or induced mutations. As Andrew Cockburn notes, "enormous random changes can result from chemical and irradiation mutagenesis, which is also traditionally used for crop breeding."[1] As already mentioned, there is no database of breeding methods that compiles adverse, unintended effects. In both traditional and molecular breeding, there is a considerable knowledge gap.

Skeptics of the mainstream position claim that molecular breeding presents unique risks because it bypasses the plant's natural gene regulation system and reprograms part of the genetic functions of the plant genome. In contrast, natural breeding uses the genes in the plants or in closely related plants and does not involve introducing foreign markers and promoter genes into a host plant genome. The skeptics acknowledge that there are risks of mutational breeding, which induces DNA mutations in plant cells (genome) by radiation or chemicals. But with mutagenesis, the germ plasm of the plant is the starting material. You can only go so far in creating mutations in the existing DNA of a plant. In molecular breeding, the scientist creates a gene cassette made up of a foreign *promoter gene*—the gene from another species and possibly kingdom that provides the plant with a qualitatively new phenotype (insect resistance), a marker gene, and a terminator gene that tells the cell when to stop transcription (when to stop the reading frame). Each of these components comes potentially from a different source that is not available to a traditional breeder.

There are also cases where molecular breeders use promoters from the parental crop or from sexually compatible plants and remove the marker genes before entering the market. A genetically engineered potato was made using genetic components from wild relatives. The genetically modified Innate potato, developed by the J. R. Simplot Company, was approved by the U.S. Department of Agriculture (USDA) in 2014 and the Federal Drug

Administration (FDA) in 2015. Innate was designed with lower amounts of the amino acid asparagine, which turns into acrylamide (a toxin) during frying. It does not contain genetic components from other species.[2]

Some commentators compare radiation-induced mutagenesis with molecular breeding, arguing that since the former is considered part of conventional breeding, and it is unregulated and unpredictable, then molecular breeding also should be unregulated. GMO skeptics like John Fagan, Michael Antoniou, and Claire Robinson argue that "Comparing genetic engineering with radiation-induced mutagenesis and concluding it is safe is like comparing a game of Russian Roulette played with one type of gun with a game of Russian Roulette played with another type of gun. Neither is safe."[3]

According to the Council of Europe in 1990, GMOs are "organisms in which the genetic material has been altered in a way that does not occur naturally by mating or natural recombination."[4] The Council excludes from its meaning of *GMO* (what I have termed *molecular breeding*) mutagenesis, cell fusion, and protoplast fusion, which they classify under traditional breeding methods. The chemical mutagenesis and radiation mutagenesis of plants affect genes, so they can fit under the strict definition of molecular breeding. However, they remain part of traditional breeding both in Europe and the United States because they do not use in vitro methods of cutting and splicing genes.

Contested Point 2: Molecular Breeding *Is* (*Is Not*) More Precise Than Traditional Breeding and Consequently *Safer* (*Not Safer*).

It is often argued that GMOs are the result of methods that use greater precision than those of traditional breeding techniques in creating genetic changes in plant species. In some respects, this is correct. Plant scientists can isolate a gene and create a specific gene cassette that is transferred to plant cells. In other forms of breeding, getting the right genes into the target cultivar is a trial and error process, and the breeding process combines more than the genes responsible for the traits.

GMO skeptics argue that precision is more than isolating a gene for transfer into plant cells. Precision also involves achieving a desired outcome from the transfer of the gene cassette into the parental plant. It is generally agreed that the gene cassette, which is introduced into plant cells by either

Agrobacterium tumefaciens or biolistics, is randomly placed in the chromosome.[5] The precision of the gene splicing is turned into imprecision when the gene cassette gets inserted into plant cells. The *position effect* (the variation of expression exhibited by identical transgenes that are inserted into different regions of a genome) on plant phenotypes is a well-recognized phenomenon. For GMO skeptics, it means uncertain risks. For GMO promoters, it means that breeders can select against any *adverse pleotropic effects* (the effects from one gene influencing two or more other phenotypic traits). Not only is there imprecision in the location of the transgene, but there is also uncertainty about whether the promoters in the gene cassette will affect other genes near or around the cassette.

In their analysis of transgenic insertions, Johnathan R. Lathan, Allison K. Wilson, and Ricarda A. Steinbrecher write: "Despite the supposed precision of genetic engineering, it is common knowledge that large numbers of individual transgenic plants must be produced in order to obtain one or a few plants that express the desired trait in an otherwise normal plant."[6] Others, like Nina Federoff and Nancy Marie Brown, are more optimistic about the plant's ability to accommodate to insertion mutagenesis or other perturbations in the plant genome: "A gene's location, while not unimportant, is less important than what the gene is and what it codes for."[7] Although the position effect is generally accepted, advocates for not treating molecular breeding as unique say that experience allows the plant breeders to select healthy plantlets with the desired traits. The GMO skeptics recoil at relying on this confidence in the breeders: "Even after selection, there are many reports of apparently normal transgenic plants exhibiting aberrant behavioural or biochemical characteristic upon further analysis."[8] H. A. Kuiper, G. A. Kleter, P. Hub, J. M. Noteborn, and E. J. Ko have documented a number of unintended effects in genetically engineered crops.[9]

Federoff and Brown believe that these aberrant effects will be picked up by the breeders' discretion before the crops are marketed: "And if an insertion is bad for the plant, the contemporary breeder does precisely what [Luther] Burbank did—throws the plant out. But modern breeders…can analyze the proteins, nucleic acids, fats and starches, as well as all of the many molecules (called secondary metabolites) that the genetically modified plant makes. They can compare this analysis to the chemical profile of the variety from which the plant was derived and ask, is this the same plant, except for that one protein encoded by the added gene."[10]

The question remains, Will the breeders perform the functions outlined by Federoff and Brown? And will an unanticipated adverse effect manifest itself right away or gradually over a period of consumption years? In the former case, stalwart GMO promoters argue that the time, effort, and cost of identifying adverse effects before commercialization are not justified, whereas GMO skeptics are faithful to the precautionary principle. In the latter case, how many years will it take for the adverse effect to show up after the product is approved and placed on the market? Can the adverse effect be picked up by premarket testing through compositional analysis or animal feeding studies?

Contested Point 3: The Concept of Substantial Equivalence *Is* (*Is Not*) an Effective Sorting Method for Determining Which Transgenic Plants Need Greater Regulatory Oversight.

The concept of *substantial equivalence* was created with two points in mind. First, it serves as a standard for sorting the products of molecular breeding into those that should and those that should not require greater regulatory oversight and risk assessment beyond what is required of the non-GM parental strain or any conventionally bred counterpart. Second, it was thought of as a starting point in risk assessment "that provides guidance by helping to identify questions to be asked during the safety assessment."[11]

The concept was first introduced by a panel of scientists who were assigned to develop safety criteria for foods and crops that were developed through biotechnology under the auspices of the Organisation for Economic Co-operation and Development (OECD), an intergovernmental economic organization. The panel's report, titled *Agricultural Policies in OECD Countries: Monitoring and Evaluation 2000*, discusses "substantial equivalence" for new foods or food components derived by modern biotechnology.[12] According to the panel's viewpoint, a transgenic crop can be compared to a parental (or conventionally bred counterpart) crop from which it was derived, about which there is extensive knowledge of possible toxicants, critical nutrients, and other relevant characteristics. When extensively analyzed, if the transgenic crop exhibits no significant changes in its components or traits compared to the parental strain (or conventionally bred counterpart), it can be treated as "substantially equivalent" to that strain. After that determination is made, further safety or nutritional concerns are expected to be insignificant. Food that is found to be substantially equivalent to a parental strain can be treated, for food safety purposes, in the same manner as its conventional counterparts.

In U.S. policy, the finding of substantial equivalence for a transgenic crop is analogous to the designation of generally regarded as safe (GRAS) for food additives. According to biotechnology promoters, the value of this category is that it prevents redundant testing and regulatory burden on producers. Unlike European policy, a GRAS finding in the U.S. policy is attributed to all GMOs prior to testing.

GMO skeptics deride the concept of substantial equivalence as vague and prone to regulatory discretion and producer lobbying because there are no canonical tests that establish equivalence in nutrients and traits. As Erik Millstone, Eric Brunner, and Sue Mayer write in *Nature*: "The concept of substantial equivalence has never been properly defined; the degree of difference between a natural food and its GM alternative before its 'substance' ceases to be acceptably 'equivalent' is not defined anywhere. … Substantial equivalence is a pseudo-scientific concept because it is a commercial and political judgment masquerading as if it were scientific … created primarily to provide an excuse for not requiring biochemical or toxicological tests."[13] Cockburn, a Monsanto scientist, disputes Millstone and his colleagues in a single sentence: "This argument which aligns with the inability to prove a negative is equally applicable to so-called conventionally bred crops and is therefore disproportionate and specious: it makes no sense to single out GM crops."[14] However, after the fact that traditional breeding and molecular breeding are distinct is accepted, then the concept of whether "substantial equivalence" is properly defined can be discussed.

What evidence is there that GMOs, which were designated as substantially equivalent, have nutritional and toxicological profiles that are not significantly different than that of the parental strains or a conventionally bred counterpart? Fagan, Antoniou, and Robinson report a number of cases where significant differences were found: "Commercialized MON810 GM maize had a markedly different profile in the types of proteins it contained compared with the non-GM counterpart when grown under the same conditions. These unexpected compositional differences also showed that the MON810 maize was not substantially equivalent to the non-GM isogenic comparator even though worldwide regulatory approvals of this maize had assumed that it was."[15] It is possible that many approved transgenic crops are not substantially similar to the non-GMO parental strain or a conventionally bred counterpart, but that does not mean they are unsafe to eat or unhealthy for the environment. The sorting criteria, the risk assessment, and the tested effects of GMOs on human consumers are all distinct issues that can be evaluated on different empirical evidence.

Contested Point 4: A Transgene That Is Used to Alter a Biochemical Pathway in a Plant, That Alters a Protein Synthesis, and That Is Not Found in Humans *Is* (or *Is Not*) Inherently Safe for Humans.

Some plant geneticists have stated that herbicides or pesticides that interact with a plant's enzyme systems and result in plant death do not affect humans who lack the same enzyme pathway. Thus, the herbicide glyphosate kills plants by interfering with the enzyme excitatory postsynaptic potential (EPSP) synthase that is essential for synthesizing certain essential amino acids (the aromatic amino acids phenylalanine, tyrosine, and tryptophan). Paul Lurquin writes that "humans do not possess the EPSP synthase, the enzyme, and as a result, our protein synthesis mechanism cannot be inhibited by glyphosate."[16] Monsanto advertises that Roundup is safe for humans and pets because it targets an enzyme that is not present in humans or pets. Athenex Corporation submitted a toxicological review to the FDA of a glyphosate tolerant enzyme (EPSPS) on behalf of a commercial sponsor: "Because the shikimate pathway [a seven-step metabolic route that is used by bacteria, fungi, algae, parasites, and plants for the biosynthesis of aromatic amino acids (phenylalanine, tyrosine, and tryptophan)] is not present in animals, glyphosate has a favorable toxicology profile and has become a very common non-selective herbicide."[17] A similar conclusion was drawn by Lurquin in his book *High Tech Harvest: Understanding Genetically Modified Food Plants* about a pathway for making plants tolerant to the herbicide glufosinate: "Glufosinate is safe for humans because we do not convert nitrate into ammonia and do not use ammonia to make the amino acid glutamine," and because other animals do not use ammonia to make glutamine, glufosinate is not harmful to them.[18]

But this leaves the question open: Can a chemical affect different pathways in plants and animals? When studying the toxicology of chemicals, carcinogenicity, genotoxicity (mutagenicity), allergenicity, endocrine disruption, and neurotoxicity must be considered. Each of these outcomes has a different pathway of action. If a chemical can have multiple endpoints of activity, why should we expect that it operates only within a single chemical pathway? This raises an important question: because we cannot know in advance how many pathways a chemical can use, how can we exercise a satisfactory risk assessment of the chemical in a new biological environment?

Cockburn addresses the issue of when a modified gene leads to a changed metabolic pathway or a new biochemical pathway. He proposes a "full analysis of the gene for open reading frames, ribosome binding sites.... Moreover, the metabolic economy of the cell may be altered upstream or downstream of the targeted change in the pathway affecting the overall nutritional and or toxicological profile of the crop."[19] This point about unanticipated metabolic pathways in GMOs was not made by a GMO skeptic but rather by a scientist at Monsanto. The issue has to be addressed by testing and not by assuming the transgenic crop is GRAS. Thus, glyphosate should not be assumed to disrupt only an enzyme pathway in plants that is not found in humans. But unless that is the only pathway with which it can interact, its other potential effects cannot be assumed not to exist.

These concerns about alternative pathways are illustrated in a 2018 study conducted by the Ramazzini Institute of Italy. Researchers fed levels of glyphosate, considered safe under U.S. standards, to rats and found that the glyphosate-fed rat pups had significant alterations in their microbiome compared to controls. This is an example of another pathway to illness through alteration of the microbiome. The authors conclude, "these data strongly indicate that GBHs [glyphosate-based herbicides] can exert long-lasting health effects later in life."[20]

Contested Point 5: GMOs *Have* (*Have Not*) Been Adequately Evaluated for Health Effects.

The scientists who have confidence in the safety of GMOs for human consumption support their claims with specific evidence or background information. A summary of these arguments follows. First, GMOs have been rigorously tested by companies and overseen by agencies to ensure their safety. These claims have been echoed in the scientific literature. For example, Suzie Key, Julian K.-C. Ma, and Pascal M. W. Drake write in the *Journal of the Royal Society of Medicine* that "GM plants undergo extensive safety testing prior to commercialization."[21]

Second, hundreds of millions of people have consumed GMOs for over twenty years with no evidence of ill effects and no lawsuits against the GMO manufacturers, even in the United States, which is a litigious nation. If GMOs are a health threat, surely we would have heard by now. Many lawsuits have been filed for the herbicide Roundup but not specifically for the health effects of transgenic crops.[22]

Third, it has been claimed that the scientific evidence is overwhelmingly in favor of the safety of GMOs in human consumption. In one widely cited review of 1,700 studies, the authors[23] state that GMOs are established as safe.[24] Another review funded by the agricultural biotechnology industry asserts "the improbability of de novo generation of toxic substances in crop plants using genetic engineering practices."[25]

Fourth, any hypothetical risks conjured up by GMO skeptics have not borne out, and any GMO crops that have exhibited significant deficiencies during the breeding process have been rejected by breeders before commercialization.[26]

Fifth, agencies like the European Food and Safety Authority (EFSA) and the U.S. Food and Drug Administration (FDA) have given GMOs currently on the market a positive safety rating. Society depends on these agencies for many other safety determinations, so why should we not respect their decisions on GMOs? Katy L. Johnson, Alan F. Raybould, Malcolm D. Hudson, and Guy M. Poppy state that "GM crops today in Europe are subject to intense scrutiny and European regulations to assess and address the safety of GM crops with regards to human health and the environment are perhaps the most stringent in the world."[27]

Regarding the first argument, GMO skeptics believe that testing done by corporate molecular breeders is largely restricted to the phenotypes with the intended or desired traits for the GM crop, such as agricultural performance. Other tests (for example, those done on crop nutritional components, allergenicity, and possible emergent unanticipated properties) are not fully transparent, often are labeled as confidential business information, and do not report negative findings. Independent toxicologists have found it difficult to obtain isogenic varieties (pure lines with and without the transgene) from seed manufacturers to undertake their own studies.

As for as how rigorous the studies are, at least in the United States, there are no required testing requirements under the 1992 FDA policy. The agency undertakes a "consultation" with a molecular breeder. The FDA classified the new GMOs as generally regarded as safe (GRAS)—a status that is assigned to certain chemical food additives. In its 1992 policy, the FDA wrote: "The agency is not aware of any information showing that foods derived of the new methods [gene splicing] differ from any other foods in any meaningful or uniform way, or that as a class, foods developed by the new techniques present any different or greater safety concern than foods developed by traditional plant breeding."[28]

Critics of this policy note that in 1992 little was known about testing GMOs for toxicity. The first products were not commercialized until 1996. So how could FDA know four years before the first products were commercialized that the foreign genetic materials introduced into plants, at random locations in the chromosomes, were GRAS? Moreover, long-term tests of GMOs on animals were not begun until over a decade after the policy was introduced.

The second argument used as evidence of the safety of GMOs for human consumption is that countless people have consumed GMOs for over twenty years with no evidence that any person has been harmed. This is certainly true. There have been no public health advisory warnings, no clinical studies documenting human health effects, and no anecdotal cases of acute toxicity from GMO crops related to the transgenes.

GMO skeptics acknowledge the lack of evidence for acute toxicity in humans. They question the safety of GMOs for chronic toxicity, where testing has been minimal. Fagan, Antoniou, and Robinson write: "Short-term studies are useful for ruling out acute toxicity, but do not provide valid evidence regarding the long-term safety of GMOs. Effects that take a long time to show up, such as cancer, severe organ damage, compromised reproductive capacity, teratogenicity, and premature death, can be reliably detected only in long-term and multigenerational studies."[29]

The GMO skeptics also question whether the assessments that are carried out by the seed manufacturers examine the subtle changes in proteins, including nutrients. Consumers might be getting a generally safe food crop with diminished nutritional value, which in the long run could affect human health.

There is a point to be made here. "Junk food," which is characterized as food that is low in nutritional value and high in calorie content, generally does not exhibit acute effects on consumers but over the long run can be detrimental to their health. The GMO skeptics offer a cornucopia of hypotheticals, including the impact of some GMO proteins (such as the Bt toxins) that show up intact in the human gut. Can they affect the human microbiome adversely? Critics call for assessments of GMOs that investigate the nutritional comparability of the GMO crop with its parental strain. Is there any scientific basis for concern? Although there are not many studies, at least one solid example gives the hypothetical risks some credence. In 1995, Tomoko Inose and Kousaku Murata of the department of applied microbiology at the Research Institute for Food Sciences at Kyoto University showed how a GM yeast exhibited increased toxicity: "The cellular level of methylglyoxal (MG), a highly toxic 2-oxoaldehyde, in *Saccharomyces cerevisiae* cells transformed with genes…was compared with that in non-transformed control cells.…When these transformed cells were used for alcohol fermentation from glucose, they accumulated MG in cells at a level sufficient to induce mutagenicity. These results illustrate that careful thought should be given to the potential metabolic products and their safety when a genetically engineered yeast is applied to food-related fermentation processes."[30]

Regarding the GMO proponents' point 5—that the scientific studies are overwhelmingly in favor of the safety of GMOs—they are correct. There are more studies that support the null hypothesis—no adverse effects.

GMO skeptics argue that a significant number of the no-effect studies were supported by industry and were based on acute toxicity studies. They were not testing worst-case scenarios and accept as an assumption "substantial equivalence." When industrial sources analyze the data of studies that report adverse effects of GMOs, they claim that the experiments possess insufficient power (that is, too few animals) to show an effect.

The twentieth-century philosopher Karl Popper noted that falsifying a hypothesis (by using deduction) can yield a definitive result but that confirming a hypothesis (by using induction) cannot. Thus, trying to prove a food is safe is always indeterminate if a case is built by accumulating supporting evidence. If the hypothesis is "The food is safe," it can be falsified by one piece of reliable evidence—that is, the food caused allergenic reactions in consumers. Even so, the food might be safe for all people who do not have allergies to it. If the hypothesis "The food is not safe" is accepted, then tests needs to be done to falsify the hypothesis and demonstrate that it is safe. Finding that it is allergenic for one individual does not mean that it is allergenic for many. And that is where population studies enter.

How does this translate into toxicology studies? The best evidence that a food is safe is testing the most probable case where it might not be. If it is believed the GMO food is unsafe because it contains the upregulation of a toxicant, then the most sensitive test for the toxin's effect should be chosen. If the test results are negative, then this is the best evidence to falsify the hypothesis "The food has a dangerous amount of a toxin." If, however, a test is underpowered or has a low probability of yielding a positive result, then it produces a weak evidentiary claim that there is no risk.

If the starting point is the hypothesis that "GMOs are safe" (that is, substantial equivalence), then what tests would most likely reveal any hazards? Applying the Popperian method of falsification requires making an effort to falsify the hypothesis. Only if it survives the test can it be assumed to be safe. The GMO skeptics argue that this method is not the one currently being used in the United States.

The fourth point made in favor of GMO safety is that plant breeders have considerable experience in eliminating unsafe cultivars before they are introduced into the marketplace. GMO skeptics argue that molecular breeding introduces new possibilities that cannot be observed by plant breeders because the effects might be hidden or subtle, such as changes in a plant's nutritional properties. Moreover, they ask whether society should place its trust in the manufacturers of a product as a definitive source of its safety. That, they say, would be a conflict of interest, like giving General Motors the final word on the safety of its cars.

Some GMO skeptics outline the tests that would provide public confidence in the transgenic crops: "A full range of 'omics' molecular profiling analysis should be carried out ... using high-throughput methods, such as microarray analysis, proteomics (study of all proteins produced), and metabolomics (the small molecule metabolites in biological systems) ... these 'omics' profiling tests must be done on the GMOs, and the isogenic non-GMOs grown at the same location and time, in order to highlight the presence of potential toxins, allergens, and compositional/nutritional disturbances caused by the GM transformation."[31] This view has been supported by the 2016 report of the National Academies of Sciences, Engineering, and Medicine (NASEM) and is discussed in chapter 11.

The FDA has acknowledged some of the possible outcomes of transgenic plants:

Additionally, plants, like other organisms, have metabolic pathways that no longer function because of mutations that occurred during evolution. Products or intermediates of some of these pathways may include toxicants. In rare cases, such silent pathways may be activated by the introduction or rearrangement of regulatory elements, or the inactivation of repressor genes by point mutations, insertional mutations, or chromosomal rearrangements. Similarly, toxicants ordinarily produced at low concentrations in a plant may be produced at higher levels in a new variety as a result of such occurrences. However, the likelihood of such events occurring in food plants with a long history of safe use is low. The potential of plant breeding to activate or upregulate pathways synthesizing toxicants has been effectively managed by sound agricultural practices, as evidenced by the fact that varieties with unacceptably high levels of toxicants have rarely been marketed.[32]

The operative phrase is "the likelihood of such events occurring in [GMO] food plants with a long history of safe use is low." Why is this so?

Both GMO critics and noncritics agree that untoward effects can occur with genetically engineered crops. However, noncritics consider that they are unlikely to occur and that if they do occur, they will be eliminated by the breeder. GMO skeptics do not accept the low-probability and low-impact event and thus call for extensive premarket testing: "What are needed are long-term and multigenerational studies on GMOs to see if the changes found in short- and medium-term studies, which are suggestive of harmful health effects, develop into serious diseases, premature death, or reproductive or developmental effects."[33]

The final argument for GMO safety is that major agencies of Europe and the United States have approved the safety of GMOs. It is certainly correct that the FDA and the EFSA have unequivocally approved GMOs as safe for human consumption. Some scientific observers, like Johnson, Raybould, Hudson, and Poppy, have stated that "GM crops today in Europe are subject to intense scrutiny and European regulations to assess and address the safety of GM crops with regard to human health and the environment are perhaps the most stringent in the world."[34]

GMO skeptics question the independence of these agencies. When some EPA staff scientists opposed approval of GMOs as GRAS without required testing with regard to human health and environmental impacts, they were overruled. Also, EFSA has been heavily criticized, say critics, for the conflicts of interest of its personnel and therefore is not a neutral agency separated from political pressures.

This chapter has reviewed opposing arguments within the scientific community on the human health effects of GMOs. There is general consensus that no GMO product has had an acute effect on consumers but disagreement on whether chronic effects and nutritional changes in genetically engineered foods are possible in current or future products and on whether whole-food animal studies can reveal adverse human health effects. And there is some consensus among different stakeholder groups on the value or feasibility of omics analyses of GE crops for reducing the uncertainty of unanticipated effects.

10   Labeling GMOs

As previously discussed, viewpoints about the health, safety, and social value of GMOs fall into contested camps. Opposing positions have not been resolved with the addition of new studies and technological innovations, like CRISPR/Cas9. The pro-GMO camps ignore or dismiss as inadequate any animal feeding studies that find adverse effects, and the anti-GMO skeptics ignore the safety assessments of scientific and regulatory bodies that find no adverse effects. Both groups seem to live in parallel universes.

Meanwhile, regulatory standards have been eased in Europe for importing GMOs, and the United States produces most of its maize, soybean, and canola with genetically modified seeds and has GMO apples on the market. Maintaining non-GMO and organic farms sufficiently distant from GMO farms to prevent GMO pollen drift and segmenting genetically modified (GM) from non-GM products in distribution networks have been formidable challenges.

U.S. citizen activism began with direct protests at supermarkets to persuade them not to stock GM produce and processed food and turned into a movement demanding mandatory labeling on those products. Food manufacturers were concerned that labeling GMOs would reduce sales. Drawing on the popular idea of consumer autonomy, the labeling movement brought support from consumers, corporate executives, and politicians who were not necessarily opposed to GMOs but who supported food choice. An op-ed in the *New York Times* was titled "I Run a GMO Company—and I Support GMO Labeling."[1]

FDA Labeling Policy

The resolution of the labeling question within the U.S. Food and Drug Administration (FDA) began with its 1992 policy statement:

> The agency is not aware of any information showing that foods derived by these new methods differ from other foods in any meaningful or uniform way, or that, as a class, foods developed by the new techniques present any different or greater safety concern than foods developed by traditional plant breeding. For this reason, the agency does not believe that the method of development of a new plant variety (including the use of new techniques including recombinant DNA techniques) is normally material information within the meaning of 21 U.S.C. 321(n) and would not usually be required to be disclosed in labeling for the food.[2]

In 2001, the FDA reviewed its policy on genetically engineered (GE) crops and reaffirmed that transferred genetic material can be presumed to be generally regarded as safe (GRAS).[3] The FDA adopted a "product" approach in contrast to a "process" approach to genetically engineered foods. It argued that GE crops and traditionally bred crops were not materially different. The agency adopted the term *substantial equivalence* to emphasize the material similitude between traditionally bred and molecular-bred crops. According to this view, there were no inherent risks of GE over non-GE food. Because crops produced through traditional breeding were not evaluated prior to entering the market, the logic of substantial equivalence would have the FDA treat non-GE and GE crops the same way. The FDA's exception to the general rule would be if a known or suspected hazard is associated with the added protein to the GE crop. Thus, if there is evidence that a new protein added to a transgenic crop is allergenic to certain consumers, FDA could require testing and labeling of the GE crop and related food items derived from it.

Under current U.S. policy, labeling of food can be mandated for a crop or processed food derived from it only if that crop is materially different from a parental variety (conventionally bred counterpart) in nutritional quality or food safety. In *Alliance for Bio-Integrity et al. v. Shalala*, the U.S. District Court found that the FDA had no basis under federal law for requiring labeling because FDA had determined that crops produced with genetic engineering posed no different or greater food safety concerns than crops produced through traditional plant breeding methods.[4]

The U.S. policy on labeling is strikingly at odds with that of the European Union (EU). The European Parliament adopted a resolution in 1997 requiring that all foods produced as or derived from GMOs be labeled as such. Subsequently, the EU expanded the rules to require labeling whenever GMO ingredients constitute at least 0.9 percent of the product. Currently, sixty-four countries (including Japan, Russia, and China) have mandatory labeling of GMO products. The European Union, unlike the United States, considers agricultural biotechnology a novel food-producing process that requires new regulations.

State Labeling Initiatives

The U.S. labeling movement began with a few state citizen petition initiatives in the states of California, Colorado, and Washington. From 2014 to 2016, around seventy GMO labeling bills were introduced in twenty-six states, and three states passed GMO labeling laws. In December 2013, Connecticut became the first state to enact labeling legislation,[5] with a caveat that the law would not go into effect unless four additional northeastern states with an aggregate population of 20 million enacted mandatory labeling laws that were consistent with the Connecticut requirements.

In January 2014, Maine became the second state to pass mandatory labeling legislation,[6] and it, too, would not take effect until four other northeastern states passed similar legislation. Vermont enacted a GMO labeling bill that was signed into law by the governor on May 8, 2014, and that was not contingent on the passage of similar legislation by other states. The law says that "food offered for sale by a retailer after July 1, 2016 shall be labeled as produced entirely or in part from genetic engineering if it is a product: (1) offered for retail sale in Vermont; and (2) entirely or partially produced with genetic engineering."[7]

The Grocery Manufacturers Association (GMA) challenged the Vermont law as unconstitutional on First Amendment grounds of free speech. The GMA argued that it was being forced through the use of its labels to convey an opinion that it did not believe. Although this challenge of commercial speech was successful in other cases, the plaintiffs failed to strike down Vermont's GMO labeling law. The state argued that the FDA does not conduct independent studies of GMOs but uses studies often paid for by GMO manufacturers, that there is a lack of consensus on health and safety, and that environmental and religious concerns contribute to a substantial government interest that overrides the protection of commercial speech. U.S.

District Court Judge Christina Reiss ruled against the Grocery Manufacturers Association and other industry groups that were plaintiffs in the case in their request for a preliminary order to block the law from going into effect as scheduled on July 1, 2016. The judge stated that the "safety of food products, the protection of the environment, and the accommodation of religious belief and practices are all quintessential government interests," as is the "desire to promote informed consumer decision-making." But the judge did not dismiss the case, giving the plaintiffs an opportunity to go to trial.[8]

Federal Labeling Bills

At the federal level, a bill was introduced into Congress that tried to preempt states from enacting labeling legislation for GMOs. The Safe and Accurate Food Labeling Act of 2015 (H.R. 1599) amends the Federal Food, Drug, and Cosmetic Act such that

> beginning on the date of the enactment of this Act, no State or political subdivision of a State may directly or indirectly establish under any authority or continue in effect as to any food in interstate commerce any requirement with respect to the sale or offering for sale in interstate commerce of a genetically engineered plant for use or application in food that is not identical to the requirement of section 461 of the Plant Protection Act.[9]

The primary motivation behind the bill was to prevent the growing patchwork of labeling laws passed by the states. The House bill would have established a voluntary certification program for genetically modified food within the U.S. Department of Agriculture (USDA), giving the agency oversight in an area usually left to the FDA. On July 23, 2015, the bill was passed by the House of Representatives by 275 to 150, but it failed to be passed by the Senate in March 2016. The Senate introduced its version of the bill (S.764) as the National Bioengineered Food Disclosure Standard, which was passed by the Senate on June 23, 2016, and signed into law on July 29, 2016 by President Barack Obama. This bill shifted the federal government's voluntary labeling to a mandatory GMO labeling provision:

> No State or a political subdivision of a State may directly or indirectly establish under any authority or continue in effect as to any food or seed in interstate commerce any requirement relating to the labeling of whether a food (including food served in a restaurant or similar establishment) or seed is genetically engineered

(which shall include such other similar terms as determined by the Secretary of Agriculture) or was developed or produced using genetic engineering, including any requirement for claims that a food or seed is or contains an ingredient that was developed or produced using genetic engineering.[10]

The law amended the Agricultural Marketing Act of 1946 by establishing a national mandatory bioengineered food disclosure standard. It defined a bioengineered food as food that was modified in a way that could not be found in nature or through conventional breeding. There was public support for a mandatory labeling bill similar to what was passed in the European Union in 2003 under the Traceability and Labeling Regulation.[11]

The current debate over whether the labeling of GMO foods should be legally mandated has generated arguments pro and con, based on scientific, legal, economic, commercial, religious, and policy considerations. The remaining sections of this chapter examine the ways that the labeling debate over GMOs has evolved.

Other Process Food Labels

Many food labels are based on factors other than nutrition or risk. We have labels for transfats, "dolphin-free tuna," and kosher food. We have labels for organic food that is not produced by genetic engineering, by ionizing radiation, or with the use of sewage sludge and that must be grown using allowable substances listed by USDA. To receive such a label, we do not have to prove that the composition of the organic product is substantially different from the nonorganic product.

The labeling for dolphin-free and kosher is voluntary, not mandated. In July 2003, the FDA announced that it would require mandatory transfat labeling effective January 2006 because there had been a finding that transfats are unhealthy and represent a risk to consumers. But they are not sufficiently unhealthy to be banned. GMO labeling can be used by manufacturers on a voluntary basis but cannot be made mandatory under agency regulations because the food product is not deemed materially different from the non-GMO counterpart. Popular opinion for labeling does not change the legal authority of the FDA to make it a requirement.

Maize, with a genetically engineered protein from the bacterium *Bacillus thuringiensis* (Bt), which is toxic to some insects, makes the GM maize materially different from the non-GM variety. It has a new protein that the parental plant (its non-GM counterpart) does not have. But according to the FDA, even with the new protein, it is nutritionally equivalent to the non-GM maize and does not present any unacceptable risk to the consumer. Unless there are risks in the Bt corn product, such as possessing allergenic properties not found in the non-GM counterpart, labels cannot be mandated under FDA regulations.

But U.S. policy does not require testing to evaluate whether there is nutritional equivalence or whether there are long-term, chronic risks from consuming the GMOs. Without required comprehensive premarket testing, how would the consumer know whether the criteria for material equivalence and safety are satisfied?

Many professional groups, including the National Academies of Science, Engineering, and Medicine (NASEM), do not believe that there is evidence of material differences between GMOs and non-GMOs or that there are risks to consumers. Unless such evidence is presented, the U.S. regulatory sector does not feel empowered to impose mandatory labeling.

Another argument of concern is that some GMOs are created with antibiotic-resistant markers. These are proteins that, if they enter our gut bacteria, could make some of them resistant to certain classes of clinical antibiotics. This is especially concerning if the bacteria in our gut happen to be pathogenic. Antibiotic-resistant markers used in the molecular breeding of crops are considered a processing aid rather than a food additive, and FDA's labeling regulations exempt processing aids from the labeling requirements.

The one area where labeling has succeeded has little to do with the product itself. It is *point-of-origin labeling*. This type of label identifies the country of origin for the product. In *American Meat Institute v. USDA*, the plaintiffs contested a national labeling law requiring food manufacturers to label all meat with the country of origin. The U.S. Court of Appeals upheld the law and disagreed with the American Meat Institute's interpretation of the First Amendment's protection against compelled speech. Not all courts have had a consistent interpretation of commercial speech under the First Amendment, however: "Where mandatory labels are permissible ... there must be a sufficiently close relationship between the government's interests, such as a specific health or safety threat, and the label."[12]

Vermont passed a law requiring the labeling of milk that is produced with bovine growth hormone—recombinant bovine somatopropin (rbST).

In *Dairy Food Industry v. Amestoy*, the U.S. Court of Appeals for the Second Circuit struck down the Vermont labeling law, in part because consumer interest was not sufficient for supporting a substantial government interest.[13] The FDA claimed that although the use of rbST affected dairy cows, such as by increasing the rate of mastitis, it had no effect on the chemical composition of milk and did not introduce human health and safety concerns.

State GMO Labeling

Consumer groups began to propose that states should adopt their own criteria on questions of material equivalence, risk, and the precautionary principle. They argued that citizens should be able to make their own choices on whether to buy or avoid GMOs. Based on public referenda, California requires labeling on products for toxic chemicals that is not mandated by the federal government or by other states.[14] State labeling laws on toxic substances, with the exception of radioactive materials, have not been preempted by federal statutes.

Until 2016, there were no federal laws that preempted state labeling statutes for GMOs. Industry groups claim that such labels will drive up the price of products. Companies often include the state-mandated labels for *all* their products regardless of where they are shipped to avoid operating with a patchwork of labeling requirements.

When states have attempted to create labeling laws for GMOs, they usually are sued by the food industry. Courts have ruled against some labeling requirements, based on claims that corporations have a First Amendment right not to be compelled to state things about what is in their food or how their food is produced that are not justified by legally valid government interests. In other words, agribusiness used a First Amendment argument that requiring food labeling makes manufacturers "speak" against their will.[15]

Every state GMO labeling initiative put to a popular vote has lost at the ballot box. This was likely due, at least in part, to voters' concerns that labeling would increase food prices. Labeling critics also argue that labeling a food "GMO" or "GMO ingredients" does not tell consumers what, if anything, is new about the food constituents or whether the production process for GMO seeds has any effect on the food.

Many ballot initiatives have failed by the slimmest of margins and failed only after the commercial interests opposing labeling spent millions of dollars on antilabeling advertisements. The forecasted effects of labeling on food prices vary significantly according to various economic analyses. Meanwhile, national polls have shown that most consumers want to know if their food is produced by genetic engineering. Consumers are increasingly concerned about how food is produced, such as whether the food is organically grown, whether food animals are grass fed, or whether chickens are "free range." The new food movement has supported consumers' right to the information they consider relevant to highly personal decisions such as the food they eat and serve to their families. Mark Bittman, former food editor for the *New York Times*, used the labeling case to advance a food transparency revolution for all aspects of food composition and production.[16]

New Federal GMO Initiatives

People have many reasons for opposing GMOs, including sustainability, seed patents, corporate contracts that restrict the use of seeds, religious principles, and unknown risks. Whatever their reason, the question is whether consumers have a right to make a choice based on food ingredients or food production methods. In response to public advocacy, the U.S. Department of Agriculture developed a new government certification program for labeling foods that are free of GMOs. That certification program, however, was overshadowed by a GMO labeling law that was signed by President Obama in July 2016 and that amended the Agricultural Marketing Act of 1946. Public Law No. 114-216, the National Bioengineered Food Disclosure Standard, gave the responsibility for writing the regulations for the new national mandatory bioengineered food labeling standards to the Department of Agriculture. The act preempts states from passing labeling standards that are not identical to the federal standard. The idea behind this law is that it would satisfy consumers who wish to make choices based on GMO content in food just as the federal organic standards (the Organic Foods Production Act of 1990) established a program for certifying products as organic. That law took a process-oriented approach because the composition of organic foods is not considered to be different than nonorganic food.

Critics point out that if companies feel that putting a GMO label on their product will reduce sales, they will prefer no label of any kind. The federal labeling law gives companies a few options for disclosing whether their products contain genetically modified ingredients. They have three options for their food package labels—a USDA-developed symbol that indicates that the package has GMO ingredients; an electronic label in the form of a QR (Quick Response) code, which is a two-dimensional barcode that is readable with an optical device like a smartphone; or a toll-free telephone number that consumers can call for information. There is no penalty for noncompliance. The bill was opposed by organic food associations and by many consumer groups that lobbied for an easy-to-read label indicating that a food product is made with or free of GMOs.

One of the main criticisms of Public Law No. 114-216 is the unavailability of direct information: "By placing the burden on the consumer to find out GMO information by either calling a telephone number or using their mobile device to access a website, consumers will lose their own valuable time and must overcome an extra hurdle to access the GMO information they wish to procure."[17]

The law has a threshold level for the amount of genetically engineered food that is in a product before it comes into play: "Such a threshold level would significantly ignore the concerns of many individuals who wish to avoid certain GMOs or GMOs entirely because of their religious or ethical beliefs."[18] One law review commentator summed up the new GMO labeling legislation by saying that "the National Standard nullifies current GMO labeling requirements, delays disclosures for another two years, minimizes the number of products that will require labeling, and limits the number of people who will have access to the label."[19]

In the United States, citizen GMO labeling initiatives became a surrogate for mandatory testing or banning genetically engineered crops and their ingredients in processed food. The existing law, which is not likely to make GMO products transparent for many consumers, will continue to fuel opposition and sustain the quarter-century-long debate. It remains uncertain how consumers will respond to a label, whether they will spend time accessing the information it offers, and what difference a GMO label will make in their shopping behavior.

11 The 2016 National Academies Study

In March 1863, while the United States was immersed in its Civil War, President Abraham Lincoln signed legislation establishing a new type of organization. The National Academy of Sciences was chartered as a private, nonprofit institution whose members were expected to provide pro-bono technical consultation to the government. The act stated that

> The Academy shall, whenever called upon by any department of the government, investigate, examine, experiment and report upon any subject of science or art, the actual expense of such investigations, examinations, experiments, and reports to be paid from appropriations which may be made for the purpose, but the Academy shall receive no compensation whatever for any services to the Government of the United States.[1]

The National Academy began with fifty charter members. Its first president, Alexander Dallas Bache, was the great-grandson of Benjamin Franklin. Over the years, it has expanded to include National Research Council (1916), the National Academy of Engineering (1964), and the National Academy of Medicine (1970), which became the National Institute of Medicine (2015). The name given to the overarching organization became the National Academies of Sciences, Engineering, and Medicine (NASEM) in 2015.

NASEM is the most eminent nonprofit, independent scientific organization in the United States. It is a self-governing, self-perpetuating organization with over 2,000 members and 400 foreign associates. New members are elected into the organization by existing members. Other than reimbursement for travel and lodging, members do not receive compensation for their consultation. NASEM also publishes the *Proceedings of the National Academy of Sciences* (*PNAS*), one of the most prestigious international scientific publications. Like the Royal Society in the United Kingdom, the

National Academies of Sciences, Engineering, and Medicine, with its cadre of Nobel laureates and medal winners, ranks among the top scientific institutions in the world.

Between 1985 and 2016, the National Academies issued nine consensus reports on agriculture and biotechnology. The first report, titled *New Directions for Biosciences Research in Agriculture: High Reward Opportunities*, was sponsored by the U.S. Department of Agriculture (USDA) and provided recommendations for where molecular genetics can contribute to studies of animals, food, crops, plant pathogens and insect pests. Subsequent NASEM reports were released in 1987,[2] 1989,[3] 2000,[4] 2002,[5] 2004 (two reports),[6] 2010,[7] and 2016[8] and cover human health and the ecological safety of bioengineered crops.

A consistent theme that runs through the reports is that there is no evidence that foods derived from genetically engineered crops pose risks that are qualitatively different than foods produced by conventional breeding methods. Nor is there any evidence that transgenic crops and the foods derived from them are unsafe to eat. The most recent NASEM report, published in May 2016, addresses the new body of animal feeding experiments that were designed to test the health effects of GMOs, epidemiological data on human health effects covering countries with different GMO consumption patterns, and new data on environmental impacts and yields of transgenic plants.[9] Because this consensus report is the most recent and engages the widest body of literature on human health effects of GMOs, I discuss how the study covers the issues and what consensus recommendations came from the report.

The mandate given to the panel of the 2016 report was to look at the food safety, environmental, social, economic, and regulatory aspects of transgenic crops. The report defines *genetic engineering* as "the introduction of or change of DNA, RNA, or proteins manipulated by humans to effect a change in an organism's genotype or epigenome."[10] It includes in its definition new gene-editing techniques such as CRISPR/Cas9. The report distinguishes between genetic engineering and biotechnology. Under *biotechnology*, the panel includes some methods that are usually classified within traditional breeding (such as chemical and nuclear induced mutagenesis) as well as some in vitro culture techniques that enable wide crossing of plants, which does not occur naturally. By establishing these definitions and distinctions, the NASEM panel revealed its judgment that a greater continuity exists between traditional breeding and molecular breeding than is usually recognized in the regulatory or nonprofit sector.

Crop Yields of Transgenic Crops

A widely heralded claim among pro-GMO advocates is that genetically engineered crops are necessary to keep pace with the world's growing population and the declining marginal advances in crop productivity from the Green Revolution. The question that has been raised from the birth of the agricultural biotechnology industry is whether GMOs will increase farmer yield. The first crops were designed explicitly for productivity. Herbicide-resistant crops were expected to reduce the competition of food plants with weeds and thus provide crops with greater access to soil nutrients, water, sunlight, and space for crop growth. Insect-resistant crops were designed to reduce crop losses resulting from insect infestations and thus improve yields.

The 2016 NASEM report is consistent with past National Academies studies by distinguishing between potential yield and actual yield. The potential yield of a cultivar is the theoretical yield that the crop could have without any limitations of water or nutrients (such as nitrogen), without any losses to pests and disease, and with an ideal supply of environmental resources (such as sunlight, carbon dioxide, and temperature). The actual yield represents the farming conditions and limitations for achieving maximum (theoretical) yield, including insect pests, crop diseases, weeds, soil acidity, and water scarcity.

Scientists have theorized that genetically engineered crops can improve potential yield by modifying the crops' biochemical pathways in order to utilize external resources (such as sunlight through photosynthesis) more efficiently or by improving the uptake of nutrients from the soil for plant growth. The 2010 National Research Council report concluded that "GE traits for pest management have an indirect effect on yield by reducing or facilitating the reduction of crop losses."[11] The NRC reported that there have been higher yields in many cases because of more cost-effective weed control and reduced losses from insect pests but that herbicide-tolerant crops have not substantially increased yield.[12] Thus far, the advancement of improved crop yield directly through genetic engineering has been mixed. The one exception is the transgenic eucalyptus tree as a source of cellulose. The transgene causes more cellulose to be found on the cell walls of the plant, which gives the transgenic crop a significant yield boost of 20 percent.[13]

The 2016 NASEM report acknowledges that a number of factors are associated with changes in crop yield for both conventional breeding and molecular breeding that involve farming practices and environmental conditions. After reviewing all the evidence on crop yield changes over the past forty years, the report states that "the nation-wide data on maize, cotton or soybean in the United States do not show a significant signature of genetic-engineering on the rate of yield increase."[14] Although the yield of some genetically engineered crops has increased substantially since 1980, it is an open question whether GE will contribute to higher yields in the future.

With respect to *Bacillus thuringiensis* (Bt) crops, the NASEM report maintains that it is not totally clear whether transgenic crops contribute to higher yields. Other factors "could confound the estimation of the apparent yield advantage of the Bt varieties."[15] The National Academies panel debunks the overzealous and empirically deficient claims of GMO proponents that higher yield was one of the critical benefits of genetically engineered crops and was therefore integral to feeding a growing planet (a global population that the United Nations estimates will reach 9.7 billion by 2050).[16] There were a few conditions under which the panel found yield increases: "Bt crops have increased yields when insect pressure was high, but there was little evidence that the introduction of GE crops were resulting in a more rapid yearly increase in on-farm crop yields in the United States than had been seen prior to the use of GE crops."[17] Cotton yields have been stymied by the ubiquitous bollworms. One study in the Punjab province in Pakistan found that Bt cotton used fewer pesticides, produced greater yields, and brought greater profits to farmers.[18] The panel's overall assessment of GMO yields is consistent with the scientific studies that show that GMO yield increases were idiosyncratic, circumstantial, and not systematic.

Health Effects of GMOs

The 2016 NASEM report devotes a chapter to the health effects of GMOs. It addresses the seminal issues that separate pro- and anti-GMO communities, such as "the assumption that a plant's endogenous metabolism is more likely to be disrupted through the introduction of new genetic elements

via genetic engineering than via conventional breeding or normal environmental stresses on the plant."[19] The National Academies panel investigated
the scientific literature that evaluated the health effects of GMOs compared
to non-GMOs. It began its investigation by summarizing what was known
about natural toxins in food, which can result from traditional breeding.
Breeders usually select against crops with high levels of natural toxins. Some
conventionally bred varieties (such as a potato variety with high concentrations of glycoalkaloids) have been taken off the market.[20] It reported
what is widely understood and uncontroversial—namely, that "crop plants
naturally produce an array of chemicals that protect against herbivores and
pathogens. Some of these chemicals can be toxic to humans when consumed in large amounts."[21] A question that has persisted is whether GMOs
will introduce an unanticipated class of new proteins or their metabolites
from foreign gene inserts that will result in adverse health effects.

Substantial Equivalence and Food Safety

The 2016 NASEM study outlines three categories of testing for evaluating food safety—acute or chronic animal feeding toxicity testing that uses
whole foods, compositional analysis that looks at proteins and other molecules in the food, and allergenicity testing that usually is done in vitro or
on humans. If GMOs are assumed to be safe or as safe as crops traditionally
bred, then until proven otherwise, safety testing ordinarily would not be
required.

U.S. policy on GMOs depends largely on the adoption of the concept
of *substantial equivalence*. The 2016 NASEM report acknowledges that "No
simple definition of substantial equivalence is found in the regulatory literature on GE foods"[22] and that its use has been debated. The National
Academies committee affirms that "substantial equivalence" remains the
cornerstone for GMO food-safety assessment by regulatory agencies and,
without offering an argument, states as a finding that "The concept of substantial equivalence can aid in the identification of potential safety and
nutritional issues related to intended and unintended changes in GE crops
and conventionally bred crops."[23] What is missing is an independent crite-
rion for *substantial equivalence* or at least an operational definition. We shall
return to the NASEM committee's use of *omics analysis* as a surrogate for
substantial equivalence.

Animal Feeding Studies

The 2016 NASEM report discusses animal studies of whole GMOs extensively to underscore uncertainties about testing protocols. The NASEM panel asked, What are the appropriate doses? What types of special testing requirements are involved when testing whole foods compared to agricultural chemicals? What is the right number of animals to be used in a study? In animal feeding studies, what percentage of the animal diet should consist of the GMO? Should the proportion of the GMO in the animal diet reflect actual feeding practice or test hypothetical situations (such as accumulation over generations)? How many feeding days should be included in the short-term animal tests? What constitutes a long-term test, and should it include a multiple-generation test?

Because the study of GMOs has not been standardized, many of the above questions have been left unanswered. The 2016 NASEM report raises questions about the utility of whole-food animal studies: "The utility of the whole-food tests has been questioned by a number of government agencies and by industry and academic researchers … and they are not an automatic part of the regulatory requirements."[24]

The NASEM study highlights the criticisms of animal feeding tests that were made by members of the scientific community and cites the fragility of any consensus on using a particular methodology for assessing GMO food risks. One group of scientists debunked whole-food animal studies because they are not sensitive enough to detect differences and there were other types of tests to evaluate safety. Another group of scientists believed that whole-food tests could be useful if their design were appropriate and if they could be performed objectively without a commercial bias.

In the 2002 National Research Council study titled *The Environmental Effects of Transgenic Plants: The Scope and Adequacy of Regulation*, the panel says that "With few exceptions, the environmental risks that will accompany future novel plants cannot be predicted. Therefore, they should be evaluated on a case-by-case basis."[25] This finding about environmental effects might be thought to apply to human health effects, too. Yet without a standard protocol and consensus on the testing requirements, there is not likely to be agreement on the interpretation of any particular test.

The 2016 NASEM report examines a number of animal feeding studies and in the process of reviewing them raises questions about their methodology and validity, especially for those that reported adverse effects. The report gives considerable attention to the 2012 toxicology study by Gilles-Eric Séralini and his colleagues, which initially was published in and then retracted by the journal of *Food and Chemical Toxicology* and then republished in a European journal. The NASEM panel echoed the criticisms of reviews and studies from Monsanto supporters. Although Séralini and his colleagues undertook the first long-term animal whole-food study for GMOs, the panel disagreed that "this one study should lead to a general change in global procedures regarding the health effects and safety of GE crops."[26] Other researchers have also called for long-term studies as recommended by Séralini et al.

Séralini and his colleagues responded to their critics in an eight-page "Answer to Critics" in *Food and Chemical Toxicology*.[27] They noted that no guidelines exist for GMO toxicity studies in vivo, responding to critics who claimed that they had failed to satisfy the standard protocols. They chose to use the guidelines of the Organisation for Economic Co-operation and Development (OECD), which were the best available for European studies.

Séralini et al. were clear from the outset that their study was not a carcinogenesis study, which would have required fifty rats for each group, as recommended by OECD guidelines. Theirs was a chronic toxicity study, which according to the OECD required only ten rats in each group. The 2016 NASEM report focused on the tumor data because it received the most media attention. NASEM did not address the primary data from the experiment—namely, sex hormone imbalances, disabled pituitary function, disruptions of estrogen-related pathways, and the enhancement of oxidative stress. Séralini and his colleagues were clear throughout their original retracted paper and the republished study that long-term studies need to be conducted to confirm their results.

The 2016 NASEM panel acknowledged that Séralini et al. followed the OECD testing guidelines that called for ten male and ten female rats, even though the experiment was criticized by the European Food and Safety Authority (EFSA)[28] for not having enough animals to report tumor incidence. Séralini et al. responded: "The criticism of the relatively low number of rats used in our experiment relies entirely on the misconception that it is a carcinogenicity study. It was not the case, as we stated clearly in the title and the introduction."[29]

The NASEM report stated: "The current animal-testing protocols based on OECD guidelines for the testing of chemicals use small samples and

Biologically relevant effect → Power analysis → Sample size → Experiment →

Statistical analysis → Statistically significant → Biologically significant

Figure 11.1
Power analysis and biological significance

have limited statistical power; therefore, they may not detect existing differences between GE and non-GE crops or may produce statistically significant results which are not biologically meaningful."[30]

The EFSA published a paper in 2011 that outlines the distinctions between statistical significance and biological significance in animal testing experiments. The former is related to statistical concepts, and the latter is related to biological considerations. Results that turn out to be statistically significant (namely, that there is a 95 percent probability that the observed effect did not happen by chance) may not turn out to be biologically meaningful. EFSA notes that "Many researchers incorrectly conclude that any statistically significant effect is biologically relevant as it is supported by mathematics."[31] Statistical significance is calculated after the experiment. But biological relevance should be considered in the design stage of the experiment before the test is begun (see figure 11.1 for the sequence of events).

A power analysis is an analytical method for determining the size of the researchers' sample in animal testing in order to obtain statistical probability of a type II error or a false negative. It is used to estimate the chances of detecting an effect before the study is conducted to determine the sample size needed.

To carry out the power analysis, one needs to stipulate several key factors— (1) effect size, (2) standard deviation for variables with quantitative effects, (3) chosen statistical significance level, (4) chosen power, (5) alternative hypotheses, and (6) sample size. Investigators will specify the first five factors in order to determine the sample size. The power or effect size also can be calculated if the sample size is fixed.[32] A power of 80 to 90 percent is usually considered good.

A power analysis is not done if there is no clear hypothesis or no prior data to enable a power determination. Séralini and his colleagues did not have a fixed hypothesis for their two-year test. It was a broad observational study with controls: "All rats were carefully monitored for behavior, appearance, palpable tumors, infections, during the experiment, and at least 10 organs per animal were weighted and up to 34 analyzed postmortem, at the macroscopic and/or microscopic levels (Table 1)." In their response to critics, the investigators encouraged others to repeat the study for any of their observed results: "We encourage others to replicate such chronic experiments, with greater statistical power. What is now urgently required is for the burden of proof to be obtained experimentally by studies conducted independent from industry."[33]

Other than the statistical significance of an animal feeding study, like EFSA, the 2016 NASEM panel cited the importance of the biological relevance of the findings: "How large a difference is biologically relevant before designing an experiment to test a null hypothesis of no difference?" A study with statistical significance does not eliminate uncertainty. It is merely the first step in reaching a sound result. The panel wrote: "If a whole-food study with an animal finds statistically significant effects, there is obviously a need for further safety testing, but when there is a negative result, there is uncertainty as to whether there is an adverse effect on health."[34] The panel states that a positive outcome in an animal feeding study (adverse effect) does not mean there was sufficient statistical power in the test. Compare this with what EFSA wrote: "a statistically significant treatment effect may exist but be biologically irrelevant because, although statistically significant, it is smaller than the predefined biologically relevant effect size."[35]

In addition to the study by Séralini and his colleagues, the 2016 NASEM panel cited other long-term rodent studies, some of which were intergenerational. Several of the studies found statistically significant differences between GMOs and non-GMOs, which the NASEM report did not consider "biologically relevant," without providing data on the normal range of variation among non-GMO crops. The NASEM committee questioned whether the statistically significant results of a study should be dismissed without sufficient reason.

Aysun Kiliç and Mehmet Turan Akay conducted a three-generation rat study in which 20 percent of the diet was Bt maize or a non-Bt maize that otherwise was genetically similar.[36] All generations of female and male rats were fed the assigned diets, and the third-generation offspring that were fed the diets were sacrificed after 3.5 months for analysis. The authors found that "although the results obtained from this study showed minor histopathological and biochemical effects in rats fed with Bt corn, long term consumption of transgenic Bt corn throughout three generation[s] did not cause severe health concerns on rats. Therefore, long-term feeding studies with GM crops should be performed on other species collaboration with new improving technologies in order to assure their safety.[37]

The 2016 NASEM report questioned Kiliç and Akay's result, arguing that "there was no presentation of standards used for finding what would be a biologically relevant difference or what the normal range was in the measurements."[38] The NASEM report questioned whether the power of the study was sufficient and also surmised that "most if not all of the rodent studies are based on widely accepted safety evaluation protocols with fixed numbers of animals per treatment."[39]

The NASEM committee raised other areas of uncertainty, such as whether the GMO and non-GMO sources were isogenic or were grown in different or unknown locations: "These problems in design make it difficult to determine whether differences [found in studies] are due to the genetic-engineering process or GE trait or to other sources of variation in the nutritional quality of crops."[40] With respect to the animal feeding studies, the most important message of the NASEM report calls for follow-up testing, not simply dismissal of some results: "In cases in which testing produces equivocal results or tests are found to lack rigor, follow-up experimentation with trusted research protocols, personnel, and publication outlets is needed to decrease uncertainty and increase the legitimacy of regulatory decisions."[41] Although some feeding studies with pigs showed no adverse effects, the panel did report statistically significant phenotypic differences (such as in conversion efficiency from the feed) between pigs that were fed Bt maize (GMO) and those that were fed non-GMO maize. No definitive conclusions can be drawn because of limitations in the studies, but NASEM could not discount the possibility that the genetically engineered food was responsible for the differences in animal effects.

The NASEM committee placed considerable value in the long-term animal feed data collected by the USDA. Comparison of livestock health and feed conversion ratios as reported in a review by Alison L. Van Eenennaam and Andrea E. Young[42] found that no adverse effects were detected in farm animals after comparing animal health before and after the conversion to GMO feedstock.

With regard to animal feeding studies, the NASEM report recommended that such studies should be conducted with consideration for statistics that are biologically relevant, with a power analysis done for each characteristic (endpoint) and follow-up studies carried out when there are equivocal results on the health effects of GMO crops. The NASEM committee called for public funding in the United States for independent follow-up studies whenever there were equivocal results.[43] Follow-up studies were rarely, if ever, conducted when there were adverse outcomes.

The Composition of GMOs Compared to Non-GMOs

Beyond animal studies, a second area of interest for evaluating genetically engineered crops is compositional analysis. Seed manufacturers can submit a compositional analysis of a new genetically modified cultivar to the FDA, but they are not required to do so. When they do submit the analysis, they select certain components of the GM crop (nutrients, antinutrients, toxicants), and the levels of expression of those components are measured and compared with similar components in the parental (or counterpart) variety from which the GM crop was developed. The analysis may report statistically significant differences between the level of the proteins in the GM and non-GM plants. This could be important if the GM plant had higher or lower levels of a toxicant or nutrient compared to its non-GMO parental or counterpart strain. Such a finding might suggest that the GMO is not materially equivalent to its isogenic non-GM strain.

But the 2016 NASEM report says that such a finding does not tell the whole story. The GMO protein may have a mean concentration of a protein that is 70 percent of what the non-GM plant yields. What is left out is the variation of the protein across a wide range of non-GM plants. Data from a wide variety of non-GM plants for the protein show a range of concentrations for each protein. If the 30 percent difference in the test samples is within the range of the natural variations that occur within the non-GM cultivar (under similar environmental conditions), then the differences found in comparing GMO and non-GMO cultivars, although statistically significant, are not biologically significant. Thus, the 2016 NASEM report states that statistical significance in compositional analysis may not reveal an important difference in test samples unless one knows the natural variations in a protein for the cultivar.

There is another caveat in trying to understand the differences in the compositional analysis: "It is difficult to know how much of the variance and range in values for the components is due to the crop variety, the growing conditions, and specific laboratory experimental equipment."[44] Head-to-head compositional studies for GMO and non-GMO cultivars could be done on isogenic crops grown side by side, and the results could reveal any compositional differences arising from the genetic engineering method. The panel also questioned whether the proteins that were selected for compositional analysis were the right choice. The current compositional analyses have not assessed whether the components measured are the appropriate ones to examine or whether differences found in measured components are indicators that there are differences in other unmeasured components. This suggests that the seed manufacturers should not choose the components for valid compositional analysis because there may be a selection bias.

After outlining the limitations in the current compositional analyses, the NASEM committee praises the importance of newly developed omics methods for understanding whether compositional changes occur in GMOs. These methods are not currently required to be used by regulatory authorities. Omics is a field of study in biology that examines distinct classes of molecules in cells, tissues, and organisms, including genes (genomics), RNA molecules (transcriptomics such as mRNA and tRNA), proteins (proteomics), and chemical metabolites or breakdown products (metabolomics). Some omics studies require mass spectrometry (liquid chromatography-tandem mass spectrometry), electrospidy ionization (ESI) (a technique used to produce ions using an electrospray and a high voltage applied to a liquid to create an aerosol), and electrophoresis (such as differential image gel electrophoresis, or DIGE).

The 2003 genetically engineered food guidelines of the Codex Alimentarius Commission—a body established in 1963 by the United Nations Food and Agriculture Organization and the World Health Organization to set standards for food safety and quality—includes some but not all of NASEM's recommendations for omics analysis, including an examination of new metabolites, food composition analysis, food components, and proteins. Codex is an influential organization in Europe, where testing of GE foods is required. As previously noted, in the United States GE foods are not required to be approved as safe by the FDA before entering the commercial market.

The 2016 NASEM report concludes the section on compositional analysis by stating that without omics analyses of crop composition, we are left with uncertainty over whether the GMO and the isogenic parental (or non-GMO counterpart) strain are materially equivalent. Despite these uncertainties, the committee appears confident about the safety of GMOs for consumers. There are other reasons that give the panel confidence that the current GMOs in circulation are not unsafe, even as there are some adverse effects observed from certain animal feeding studies that have not been replicated for confirmation.

Global Health Data

The 2016 NASEM report surmises that if GMOs have a significant impact on human health, the effects would show up in global health data. The committee looked at global data on cancer, chronic kidney disease, obesity, celiac disease, and allergies before and after the introduction of GMOs and compared data from countries that had heavy consumption of GMOs with countries that had light consumption. Among the data consulted was cancer incidence in women and men in the United Kingdom (a country where GMOs generally were not consumed) and in the United States (where most of the soybean and maize were GMO varieties). The committee found that the patterns of change in cancer incidence in both regions were generally similar, even though European diets contain much lower amounts of feed derived from genetically engineered crops. There was no unusual rise in cancer incidence for specific types of cancers in the United States after 1996, when GMOs were first introduced. Because of this, the report states that global cancer data "do not support the hypothesis that GE foods have resulted in a substantial increase in the incidence of cancer" from the consumption of GMOs and that the findings are clear about no findings of cancer risk.

Although a few animal feeding experiments with GMOs had cancer endpoints, cancer was not a major postulated outcome of consuming GE crops. Based on the global data, kidney disease, autism, and food allergies were not attributable to GMOs. Overall, "the committee found no differences that implicate a higher risk to human health from GE foods than from their non-GE counterparts."[45] In its final conclusion on the health effects of GMOs, the NASEM committee wrote that "the research that has been conducted in studies with animals and on chemical composition of GE food reveals no differences that would implicate a higher risk to human health from eating GE foods than from eating their non-GE counterparts. The committee could not find persuasive evidence of adverse health effects directly attributable to the consumption of GE foods."[46]

In 2015, I published a paper that examined eight systematic reviews of research on the health effects of GMOs published between 2008 and 2014.[47] The reviews differed significantly in their conclusions about what the current science concludes about the health effects in animal feeding studies. One review states that GMOs may cause hepatic, pancreatic, renal, and reproductive effects on animals,[48] and another review states that the research does not suggest any health hazards or any statistically significant differences between GMO and non-GMO crops in the parameters observed.[49] I also found twenty-six studies that individually report adverse effects of GMOs in animal feeding studies.

After reviewing the 2016 NASEM report, several questions came to mind. Does the study cite the reviews that I cover in my 2015 paper? Does it cite the twenty-six studies? Does it dismiss the reviews and studies that reported positive effects from GMOs because of methodology, weak statistical power, or lack of replication?

In the chapter on health effects of GMOs in the 2016 NASEM report, there are 287 references. That report cites four of the eight systematic reviews that I cover in my paper. Of the four that it does not cite, some report that GMOs cause adverse effects, and others reported no differences between GMOs and non-GMOs in animal studies. The NASEM report cites or acknowledges only four of the twenty-six individual studies that I found that report adverse effects. Of the studies not cited in the NASEM review, many appeared in respected journals such as *The Lancet, Journal of Anatomy, Journal of Agriculture, Food and Chemistry, Animal Science, Archives of Environmental Contamination and Toxicology, Journal of Biological Sciences, Reproductive Toxicology and Critical Reviews in Food*, and *Science and Nutrition*.

This leaves me with more questions than answers. The NASEM panel gives no specific reasons for not acknowledging four reviews and twenty-two studies. Would these studies have shifted the weight of evidence? It does not seem likely that these additional studies would have shifted the outcome of the report, given the authoritative positions cited in the 2016 NASEM report—from past National Research Council studies, the American Association for the Advancement of Science, the American Medical Association House of Delegates, the World Health Organization, the FDA, and the European Commission—that have expressed confidence in the safety of GMOs for human and animal consumption. As for the results by Séralini and his colleagues, which were retracted after a year without the authors' consent from the journal of the initial publication, the 2016 NASEM report cites the fact that no dose-response relationship was found and that the reanalysis by the European Food Safety Authority found no statistically significant differences between the GM and the non-GM crop. Séralini et al.'s work was labeled as undependable and, like most whole-food animal studies, not a credible way to assess the safety of GMOs.

The Integrity of the 2016 NASEM Study

It is now generally recognized that corporate funding of research can bias the outcome in favor of the financial interests of the funder. Most major journals have adopted financial conflict-of-interest (fCOI) disclosure policies for their contributors. Government agencies also require fCOI disclosure for members of advisory panels.

In the 1997 amendments to the Federal Advisory Committee Act (FACA), requirements regarding financial conflicts of interest were established for committees of the National Academies of Sciences, Engineering, and Medicine (then called the National Academy of Sciences). The new rules stated that federal agencies cannot utilize the scientific advice of NASEM unless two conditions are met. First, no individual appointed to serve on a NASEM committee can have a conflict of interest relevant to the functions to be performed unless NASEM determines that the conflict is unavoidable, in which case it must be promptly and publicly disclosed. Second, the committee membership must be fairly balanced.[50]

In response to the FACA requirements, NASEM developed its own conflict-of-interest guidelines. Some of the NASEM panels were criticized for having a high percentage of scientists (somewhere between 20 and 25 percent) who had ties to industry.[51] The 2016 NASEM genetically engineered crop report states that there were no conflicts of interest among the twenty scientists who comprised the panel. A February 2017 study of the panel members for the report found that six out of the twenty had financial interests in genetically engineered crops, including patents and corporate research grants.[52] The Academies dismissed the results of the 2017 study in the following news release:

The National Academies of Sciences, Engineering, and Medicine have a stringent, well-defined, and transparent conflict-of-interest policy, with which all members of this study committee complied. It is unfair and disingenuous for the authors of the PLOS article to apply their own perception of conflict of interest to our committee in place of our tested and trusted conflict-of-interest policies.[53]

In May 2017, the president of NASEM was quoted in *The Scientist* as stating that the organization would be revising its conflict-of-interest policy.[54]

In conclusion—notwithstanding the caveats expressed throughout the voluminous 2016 NASEM assessment about the uncertainties in the studies (both positive and negative), the limitations of the compositional analysis, and the importance of follow-up studies—the panel affirms that there is no evidence of health hazards from GMO food consumption. The report offers a comprehensive and nuanced analysis of genetically engineered crops, but it does not close the chapter on the health effects of GMOs. There needs to be a consensus on standardized tests (including compositional omics analysis, in vitro tests, and animal feeding studies), publicly funded research on molecular approaches for testing future products, and follow-up testing for any equivocal results. There is no evidence that the report changed the minds of the legions who remain skeptical about GMO products.

12   The Promise and Protests of Golden Rice

The first genetically modified crop that captured international attention for its potential humanitarian benefits was a nutrient-fortified transgenic cultivar of rice. Until 2000, the products developed through molecular breeding were designed for adding value to the production of agriculture or to the shelf life of agricultural produce. In the 1990s, Ingo Potrykus, a plant scientist and professor at the Swiss Federal Institute of Technology (SFIT) in Zurich, began a research program for creating rice that is enriched with provitamin A or beta-carotene. His purpose was to save millions of children with *vitamin A deficiency* (VAD) (also known as *hypovitaminosis*), a condition that has plagued children in Africa, Asia, and the South Pacific and that can result in blindness and death. Born in 1933, Potrykus was educated at the University of Cologne in plant sciences, where he earned his doctorate at the Max-Planck Institute for Plant Breeding Research.

Potrykus began his career in plant genetics in the 1970s, when he sought to transfer genes by treating *protoplasts* (plant cells without a membrane) with naked DNA (purified DNA sequence without proteins, including histones, which are the proteins on chromatin). He then turned to the use of *Agrobacterium tumefaciens* for gene transfer in tobacco and soon became interested in cereal crops such as wheat, maize, and rice. By 1990, Potrykus became acquainted with the issues around food insecurity when he learned that the major nutrient deficiencies were from iodine, iron, and vitamin A.

It is estimated that vitamin A deficiency is responsible for about a million deaths annually and that about 230 million children are at risk of subclinical VAD, which puts them at risk for infections. Blindness from VAD affects about 500,000 children per year. An estimated 3 billion people depend on rice as their staple crop, and about 10 percent of this population is affected by VAD. Vitamin A has been shown to play an essential role in

vision, immune response, epithelial cell growth, bone growth, reproduction, and embryonic development.[1]

Following recommendations from the International Food Policy Research
Institute, Potrykus decided to endow major staple crops with missing
micronutrients. His research group at SFIT and his collaborator Peter Beyer
at the University of Freiberg dedicated themselves to improving the yield
and nutritional quality of crops that were essential for developing nations,
particularly for malnourished preschool children.

Gary Toenniessen, former program officer of the Rockefeller Foundation's International Program on Rice Biotechnology, traces the origins of
the Golden Rice initiative to an international agricultural meeting in the
Philippines in 1984. Toenniessen asked a group of plant breeders how
recombinant DNA technology could benefit rice. Their answer was "yellow
endosperm." In other words, get the precursors for vitamin A into the rice
endosperm.[2] In 1992, scientists meeting in New York City began discussing ideas for introducing the provitamin A synthesis pathway into the rice
endosperm and for determining the promoters that they might use. Both
Potrykus and Beyer were at the meeting. Beyer was working on the regulation of the terpenoid pathway in daffodils. Terpenoids or terpenes are the
largest class of chemicals produced by plants, which include those responsible for smell and color. Beta-carotene is a terpenoid compound responsible
for the orange color of carrots.[3]

The SFIT team had been investigating the pathways for vitamin A production (particularly the precursors) and the possibility that they could be
brought into the rice cultivar. Unlike other products in agricultural biotechnology, where a single gene was isolated for transfer into a plant, the rice
cultivar required Potrykus to deal with multiple genes to activate a complete synthetic pathway. His work was initially supported by the Rockefeller
Foundation, known for its work in funding the Green Revolution, which
improved rice yields in Asia.

Golden Rice was promoted by the biotechnology industry as an indis-
putably humanitarian agricultural product. In parts of the world where
rice is the primary source of food but where people do not have affordable
access to foods rich with vitamin A, the risk of blindness has been compara-
tively high. In 2003, nutrition expert Marion Nestle wrote that Golden Rice
"has been industry's primary advertising tool to promote the humanitarian benefits of food biotechnology."[4] Nestle, who received her doctorate in

molecular biology before she became internationally recognized in the field of nutrition, wrote a comprehensive and balanced analysis of Golden Rice. In her book *Safe Food: The Politics of Food Safety*, which covers science, risk assessment, and politics, she writes that the creation of Golden Rice was a notable achievement in crop molecular genetics and that it could not have been accomplished by traditional breeding.[5]

The grain of rice consists of three primary components—an outer covering (husk plus bran) of nutrient proteins; an inner volume (called the *endosperm*), which consists of starch and some protein; and the embryo (a segment of the inner volume), which can grow into a plant. Like many plants, rice has small amounts of beta-carotene, a precursor of vitamin A, located in its outer layer, which is removed or milled to make white rice. There are otherwise no natural provitamin A–containing rice varieties among the rice cultivars. Rice plants produce provitamin A (beta-carotene) in green tissues but not in the endosperm (the edible part of the seed). The outer layer of the dehusked grains contains some nutrients but not provitamin A. Whatever amounts are in the husk are eliminated after milling and polishing to make white rice. The brown rice with husk removed but with bran has more nutrients, but the oils can turn rancid when exposed to air. In order to produce vitamin A–biofortified rice, the molecular breeding must transfer the vitamin A precursor synthesis protein to the endosperm portion of the rice, which most consumers eat because the husk and bran of the rice are removed.

Peter Burkhardt, who worked with Potrykus, identified one of the beta-carotene precursors as the chemical GGPP (geranylgeranyl-pyrrophosphate), which is found in the rice endosperm as well as in carrots and green vegetables. With this chemical, it was shown that beta-carotene (which has four enzymes and is one of the most common and important of the family of carotenoids for vitamin A activity in human body) could be synthesized.

Genetic engineering was needed to introduce the genes for the enzymes to activate the beta-carotene synthetic pathway into the rice endosperm. Potrykus had to develop gene constructs (gene-expression cassettes) and insert them into rice cells. One of Potrykus's students, Xudong Ye, developed a beta-carotene strain of rice and showcased his results at a 1999 symposium: "At this symposium Xudong Ye presented the results for the first time to the public. He demonstrated that he could engineer the entire biochemical pathway of β-carotene into the rice endosperm. The endosperm contained good quantities of pro-vitamin A, beautifully visible as 'golden' color of different intensity in different transgenic lines. The best pro-vitamin A line had 85% of its carotenoids as β-carotene."[6]

The gene cassette for Golden Rice contained a promoter, a terminator, and a transit sequence used to get the enzyme genes into the endosperm. The SFIT team took genes from a daffodil and a gene found in a soil bacterium (*Erwinia uredovora*) for the promoters. According to some scientists, the number of transgene constructs introduced into the rice cells contributes to its instability. Also, because the transgene constructs are randomly inserted into the rice chromosome, "each transformed cell will have its own distinct pattern of inserts."[7] *Science* published the first report announcing pro-vitamin A in the rice endosperm on January 14, 2000; its seven authors included senior scientists Beyer and Potrykus.[8]

The best strain of Golden Rice developed in 2000 delivered 1.6 micrograms of provitamin A per gram of rice grain. Critics noted that an average two-year-old would need to consume 3 kilograms of rice each day to achieve the recommended daily dose of vitamin A. This was the first-generation Golden Rice, and its inventors acknowledged that it was merely a proof of concept: "In a proof-of-concept study, we have shown that it is possible to establish a biosynthetic pathway de novo in rice endosperm, enabling the accumulation of pro-vitamin A. Many variations of this applied technology appear feasible and work is in progress to optimize the yellow rice lines now in our hands. In part, this involves the use of different structural genes and the use of different selectable marker genes."[9] Early criticisms of the low-yield GR varieties were eclipsed by later developments in higher yields of provitamin A with more preferable rice varieties.

Human Trials of Golden Rice

A group of scientists at the Jean Mayer USDA Human Nutrition Research Center on Aging at Tufts University supervised a clinical trial to test whether Golden Rice was an effective source of vitamin A. Five adult volunteers (three women and two men) from the Boston area were enrolled in the study, which lasted for thirty-six days. The trial subjects each received a single serving of Golden Rice and were monitored for any adverse effects. Investigators found no allergenic reactions or gastrointestinal abnormalities. The study, published in the *American Journal of Clinical Nutrition*, reported that "β-carotene derived from Golden Rice is effectively converted to vitamin A in humans."[10]

The same journal published a paper in August 2012 with several of the same authors reporting on the results of another clinical trial on Golden Rice. The investigators involved in the study included scientists from both the Tufts University Medical School and the Chinese Center for Disease Control and Prevention. The investigative team enrolled seventy-two healthy children between six and eight years old from Huan, China. The purpose of the trial was to evaluate the efficacy of Golden Rice compared to vitamin A supplements in getting vitamin A to young children. The children were divided into three groups—one group that was fed 60 grams of Golden Rice for twenty-one days, another group that was fed vitamin A capsules along with plain white rice, and a third group that was a control.[11]

Investigators measured the children's blood concentration of vitamin A. The results of the study showed that Golden Rice was comparable to the capsules in delivering vitamin A. On September 2015, three years after the paper was published, it was retracted by the journal editors for violation of ethical principles in the clinical trial. The informed consent form approved by Tufts University in 2008 said that the children would be given "Golden Rice" and failed to mention that the rice was genetically modified. The informed consent form allegedly did not provide parents with sufficient information about the trial.[12]

Between 1998 and 2010, the People's Republic of China passed a series of laws on clinical trials. China's laws require that its clinical investigators comply with the ethical principles in the Declaration of Helsinki, which emphasizes informed consent. The law provides that the written informed consent form must be written in language accessible to ordinary people and must clearly state the purpose and risks associated with the trial.[13] The retraction of the paper was one of several factors that slowed the progress of genetically engineered biofortified food.

Intellectual Property and Humanitarian Access

Potrykus and Beyer helped establish the Golden Rice Humanitarian Board, whose purpose was to bring the benefits of Golden Rice to developing countries at a cost no greater than the rice they currently consume. They believed that traditional breeding could be used in the Golden Rice strain to transfer the trait of beta-carotene synthesis into locally best-adapted rice cultivars.

In 2000, Potrykus and Beyer worked out an agreement with Syngenta that the company would develop the next generation of Golden Rice and that farmers who earn less than $10,000 a year would be able to get the technology for free.

Syngenta scientists replaced the daffodil gene with a maize gene that produced more than twenty-three times provitamin A (37 micrograms) per gram of rice grain than the daffodil. They estimated that 72 grams of rice per day per child would be sufficient to prevent vitamin A deficiency. According to Guangwen Tang and his colleagues, a daily intake of 60 grams of rice (about half a cup) would provide 60 percent of the Chinese recommended intake of vitamin A for children six to eight years old and would be sufficient to prevent vitamin A deficiency.[14]

The White House Office of Science and Technology Policy and the U.S. Patent and Trademark Office honored the Golden Rice project as one of the winners of the 2015 Patents for Humanity Awards.[15]

Opposition to Golden Rice

In his 2001 paper "The Golden Rice Tale," Potrykus says that "In fighting against 'Golden Rice' reaching the poor in developing countries, GMO opposition has to be held responsible for the foreseeable unnecessary death and blindness of millions of poor every year."[16] He tried to understand Greenpeace's opposition to Golden Rice and concluded that the group regarded it as a "Trojan Horse" that would allow into commerce other objectionable GMO products.

Fifteen years later, spearheaded by Richard Roberts, chief scientific officer of New England Biolabs, 107 Nobel laureates signed a letter calling on Greenpeace and its allies to end their opposition to Golden Rice:

> Scientific and regulatory agencies around the world have repeatedly and consistently found crops and foods improved through biotechnology to be as safe as, if not safer than those derived from any other method of production. There has never been a single confirmed case of a negative health outcome for humans or animals from their consumption. Their environmental impacts have been shown repeatedly to be less damaging to the environment, and a boon to global biodiversity.[17]

The Institute of Science in Society, a nonprofit organization that defines its mission as "evaluating science and technology for sustainability and social accountability," titled its 2000 essay "The 'Golden Rice': An Exercise on How Not to Do Science." The authors say that "The 'golden rice' project was a useless application, a drain on public finance, and a threat to health and biodiversity. It is being promoted to salvage a morally as well as financially bankrupt agricultural biotech industry.... it is obstructing the shift to sustainable agriculture."[18]

Delayed Development

It has been seventeen years since the first Golden Rice strain was announced in *Science* magazine. Eight years later, the same magazine wrote: "Almost a decade later, golden rice is still just that: a promise. Well-organized opposition and a thicket of regulations on transgenic crops have prevented the plant from appearing on Asian farms within 2 to 3 years as Potrykus and his colleagues once predicted."[19]

The International Rice Research Institute (IRRI) has been working on developing a Golden Rice strain since 2006. The major rice varieties grown and consumed in Asia are indica rice. Golden Rice was developed in the japonica strain. The challenge for IRRI was to cross-breed the japonica strain with the beta-carotene pathway into the indica strain and produce sufficiently high yields with locally adopted farming practices. It has reported from field trial data that the rice quality and the beta-carotene levels in the rice were comparable to traditional varieties. The problem it faced was that the yields were not consistent across locations and seasons, a limitation called *yield drag*.

Justus Wesseler and David Zilberman have developed an economic model that explains the perceived economic benefits and costs of introducing Golden Rice. The benefits are preventing blindness in children, saving lives, and improving economic development in VAD-afflicted developing countries. The costs are the costs of research and development as well as the objections of opposition groups like Greenpeace: "The opposition to GRS [Golden Rice strategy] has substantial power and indicates that it will be difficult for those supporting the technology to change the view on perceived costs."[20] The authors describe costs and uncertainties in India (such as the allegations of farmer suicides and dead sheep linked to genetically modified cotton cultivation), the adverse environmental impacts of Bt egg-plant, and the health impacts of antibiotic selection marker genes. They conclude that "A small industry has developed around the opposition to transgenic crops that survives mainly on donations and has to keep the debate about the risks of the technology alive. This strategy seems to be a successful strategy albeit, as the case of Golden Rice shows, at the cost of the lives of several thousand children."

Conclusion

The story of Golden Rice is not over. It remains a promise unfulfilled but nonetheless actively pursued.[21] The International Rice Research Institute (IRRI) has been collaborating with research agencies in Bangladesh, Indonesia, and the Philippines to fulfill the promise. IRRI and the Philippine Rice Research Institute have conducted greenhouse and field trials, and similar testing is going on in Bangladesh with its local rice varieties.

According to IRRI, Golden Rice "will only be made available broadly to farmers and consumers if it is successfully developed into rice varieties suitable for Asia, approved by national regulators, and shown to improve in community conditions."[22] IRRI's Golden Rice project was given a boost by a 2009 paper that reported a beta-carotene conversion rate from Golden Rice to vitamin A after consumption was significantly higher than common vegetables such as spinach: "After consumption, the stably [radio] labeled β-carotene from the Golden Rice was absorbed intact in the gastrointestinal tract. Subsequent conversion to vitamin A was estimated to occur at a rate of 3.8 to 1. This is much improved from the rates of 10:1 to 27:1 estimated previously for colored vegetables such as spinach and carrots."[23] These results indicate that a cup of Golden Rice eaten a day can supply 50 percent of the recommended dietary daily allowance of vitamin A.[24]

The development of Golden Rice for commercial markets in the Philippines could be hampered by the 2015 ruling of the Supreme Court of the Philippines that ordered a permanent ban on field trials of genetically modified eggplant and a temporary ban on approving applications for the contained use, import, commercialization, and propagation of GM crops. This was a victory for Greenpeace, which has actively pursued such bans. Stable cultivars of Golden Rice that have proven yields, have high conversion rates to vitamin A, and meet rigorous safety testing could change the balance of public adoption and acceptance.

The International Service for the Acquisition of Agri-biotech Applications has found that "A significant portion of the developing world's population relies largely on one or more of the staple crops for their nutrition, and these are the subject of bio-fortification projects, both by conventional breeding and by modern biotechnology methods."[25] Conventional breeding for biofortification is limited when there is insufficient variation in the crop population for enhancing a desired trait. With the apparent scientific success in the use of biotechnology for creating vitamin A–fortified rice, other crops were selected for research and development for improving nutrient composition, primarily vitamin A (beta-carotene), vitamin C, folate, iron, zinc, and proteins. The genetically modified biofortifed crops currently under development include cassava (vitamin A), potato (vitamin C), tomato (vitamin C), broccoli (vitamin C), wheat (zinc), maize (beta-carotene), and rice (folate).

Judging from the long gestation of Golden Rice's development, other biofortified genetically modified crops may face similar constraints and public skepticism. Some commentators believe that gene editing (CRISPR/Cas/9) methods will speed up adoption. According to Kathleen L. Hefferon, "It is less likely that new varieties of crops which harbor the small nucleotide modifications that are created by genome editing will be subjected to the same strict set of regulations as are currently held for transgenic crops."[26] These areas of GMO products have seen more support within the scientific literature on public health grounds than the first generation of transgenic crops.[27] Research and development of biofortified transgenic crops have grown rapidly, with provitamin A potatoes and maize leading the way.[28]

13   Science Studies and the GMO Conflict

Within the GMO debates, social scientists have found a rich trove of data to explore questions on the interactions between science and society. Beyond the technical scientific questions, social scientists explore issues such as, What factors shape the public acceptance or rejection of a new technology? What role does science play in the social narratives and policy struggles that take place both within and across national boundaries? How do we understand conflicting perspectives within the sciences? These issues drive scholarly work in science and technology studies (STS). Science studies gained traction as a discipline in the 1960s contemporaneously with the publication of Thomas Kuhn's 1962 book *The Structure of Scientific Revolutions*. Kuhn's exploration into the history of science shows that to understand revolutionary change in science, we need to look outside of science. Moreover, Kuhn denies that science is constantly approaching the truth. Rather, it builds *paradigms*—conceptual models or theories that help us understand the physical world. When one paradigm displaces another, it does not happen by scientific logic alone.

The STS tradition views science as a group of technical disciplines that engage dialectically (via a mutual interaction of different perspectives toward a synthesis) with the society in which it is embedded. Science shapes society, and society shapes science. Given the multidecade conflict over GMOs, it is not surprising that STS investigators have sought to bring clarity to the conflicting perspectives and have been called on to explain differences among scientific groups and between the science establishment and the public. Steven Yearley explains the STS interest in food biotechnology: "the case of GMOs is of wide interest to STS researchers because of the light it throws on comparative safety assessment and the interpretation of scientific evidence and precaution in international perspective."[1]

A clear expression of the STS approach toward GMOs is given by Amaranta Herrero, Fern Wickson, and Rosa Binimelis: "Researchers have typically conceived and assessed GMOs as neutral, autonomous and individualized technological objects, ignoring the socio-economic and ecological relations these technologies require, create, and or perform. In fact, as any technology, GMOs do not exist in a vacuum but operate as a socio-technical and socio-ecological systems. They inevitably affect and are affected by the networks of relations in which they both circulate and generate."[2]

In her book *Designs on Nature: Science and Democracy in Europe and the United States*, Sheila Jasanoff writes: "For politically engaged scholars of science and technology in society, following the scientists around ... provides only a tiny peephole on the power of science. It is just as necessary to chart the trajectories of the myriad other social actors whose values and expectations interpenetrate with those of scientists and inventors, creating the conditions in which scientific ideas are translated into material and social realities."[3] Jasanoff's book explores how three different political cultures (Germany, the United Kingdom, and the United States) have addressed GMOs by navigating through a common body of international science within their unique "culturally conditioned framings."

The STS scholarship has sought to understand the structural frameworks within which the GMO controversy is embedded and the ways that stakeholder interpretations of those frameworks explain the persistence of contested science and policies. Paramount among the frameworks examined are risk assessment, the political economy of GMO agriculture (including the corporate hegemonic control of GMO agriculture), and the intellectual property regime of genetically engineered crops.

Risk Assessment

The health and environmental concerns of genetically engineered crops have been a central theme among GMO skeptics. The response by GMO seed developers and federal agencies was to turn the issue into scientific risk assessment. In the United States, this has meant reframing the risk issue into a nonissue by asserting the principle of "substantial equivalence" and arguing that traditional breeding and molecular breeding have equal risks for creating unsafe crops. In contrast, the European Union has established requirements for testing whole GMO foods, setting barriers for the importation of GE crops and planting of GE seeds. STS scholarship has traditionally investigated risk assessment for a wide range of technologies. Building on the rich contributions to risk studies, the GMO scholarship has distinguished between technical and cultural concepts of risk. In the former case, risk analysis is a scientific program that is distinct from policy and values. In the latter case, risk is a highly politicized concept that cannot be separated from value judgments. In other words, under the cultural framework, risk is socially constructed. Yearley notes, "in the larger European countries there has not tended to be a single agency charged with making such [GMO] precise risk calculations. ... If risk could be objectively assessed, there should be no need for segregated responsibility."[4]

For example, Marjolein B. A. Van Asselt and Ellen Vos see GMOs as embodying "uncertain risks," as distinguished from "traditional, simple risks," which can be understood in terms of statistical techniques.[5] Skepticism among stakeholders and scientists toward GMOs is rooted in the refusal and inability of corporations and agencies to address uncertain risks leading to what the authors call "organized irresponsibility." This amounts to hiding behind "overconfidence" and ignoring "uncertainty," otherwise known as "uncertainty intolerance." Building on the wealth of scholarship on the sociology, social psychology, and anthropology of risk,[6] STS scholars classify different risk cultures and emphasize that risk analysis is intrinsically a social process.[7]

Alternatively, non-STS scholars view risk as strictly a scientific concept but as one that is vulnerable to distortions by cultural values. Ronald Herring and Robert Paarlberg try to understand why Bt cotton was accepted in India but Bt eggplant was rejected. Rather than addressing cultural differences in the standards for accepting an industrial product (cotton) in contrast to a food product (eggplant), they focus on the universality of the science of risk assessment. They argue: "Food crops are inherently susceptible to anxiety framings of importance to consumers."[8] Even without credible scientific evidence of health risk, the authors claim that critics were able to raise unsubstantiated concerns about a potential toxin in Bt eggplant, which influenced the Indian authorities to restrict the product. But as the authors note, the authorities were influenced not just by potential risks but also by "public worries about the domination of India's food supply by Monsanto."[9] This brings us to political economy as a divisive frame for GE crops.

The Political Economy of GMO Agriculture

The introduction of biotechnology to plant breeding was accompanied by a set of changes in the seed industry. These changes brought seed development more in line with industrial capitalism and the commodification of agriculture. Traditional breeding took place in laboratories and greenhouses but was laborious, time consuming, and inefficient compared to molecular breeding, which had less trial and error but was still laborious. Small seed developers were viewed as vestigial remains of a nineteenth-century production system. Neoliberal restructuring of agriculture saw a concentration in the seed industry that allowed GMO seed manufacturers to gain hegemony over U.S. agriculture.

Agricultural sociologists introduced the term *food regimes* to distinguish neoliberal from traditional food economies. A *food regime* has been defined as a "rule-governed structure of production and consumption of food on a world scale."[10] Philip McMichael notes that

> food regime analysis is key to understanding a foundational divide between environmentally catastrophic agro-industrialisation and alternative, agroecological practices that is coming to a head now as we face a historic threshold governed by peak oil, peak soil, climate change, and malnutrition of the "stuffed and starved" kind across the world. This divide is, arguably, endemic to capitalism, and its food regime at large—generating a rising skepticism regarding the ecological and health impact of industrial food, … and a gathering of food sovereignty movements across the world … to reverse the modernist narrative of smallholder obsolescence etched into the development paradigm and current development industry visions of "feeding the world.."[11]

The GMO divide has grown across competing food regimes and has distinguished traditional food economies (agro-ecological, low- or no-pesticide farming with local and regional farm distribution) from global, industrial, fossil-fuel-based agriculture.

One of the notable transformations in GMO agriculture is the changing relationship between seed developers or breeders and farmers. This is best illustrated by some changes that Monsanto instituted. Under Monsanto's technology stewardship agreement, farmers are granted a limited license to use Roundup Ready seeds. The company licenses its seeds to growers analogous to the way that Microsoft licenses its software to computer users. The seed is said to contain Monsanto technologies—namely, the transgenes and methods of constructing the GMO. Anyone who uses the seed or technology must have a licensing agreement with Monsanto. A grower cannot share the seeds with another grower. The agreement states that the grower is "Not to transfer any seed containing patented Monsanto technologies to any other person or entity for planting."[12] Growers must allow Monsanto access to their crop land to obtain any samples of crops and seeds for inspection. The grower cannot do research on the seeds or crops produced from it. Thus, for the first time in the history of agriculture, farmers do not own and fully control their seeds. They lease them from Monsanto—or so it seems. This transformation in the seed economy has contributed to the GMO opposition among traditional farmers and food activists.

Herring and Paarlberg, two astute observers of food biotechnology, emphasize the role that political economy has had on the opposition to GMO adoption. They write that "The blockage of GE food crops does not derive from scientific evidence of new risks.... [A] new structure of transnational resistance to modern farming had emerged driven by opponents of corporate-led globalization and by advocates for the environment."[13] The patenting of genetically engineered crops was part of the antiglobalization opposition.

Intellectual Property Regime

How has the STS scholarship interpreted the importance of patenting of GMOs as a key factor in public opposition? Early opposition to GMOs was ignited when patents were issued to genetically engineered seeds. In her book *Biopiracy: The Plunder of Nature and Knowledge*, anti-GMO international activist Vandana Shiva writes that "Patenting living organisms encourages two forms of violence. First, life-forms are treated as if they are mere machines, thus denying their self-organizing capacity. Second, by allowing the patenting of future generations of plants and animals, the self-reproducing capacity of living organisms is denied."[14]

The tradition of patenting in the United States is rooted in the Constitution. The Patent Act of 1790 establishes that patents will be issued to any person who has "invented or discovered any new and useful art, machine, manufacture, or composition of matter, or any improvement of these not before known or used."[15] Patenting of microorganisms grew out of that tradition "if they have been identified and isolated for the first time and shown to have practical utility."[16] For example, Louis Pasteur was awarded a U.S. patent for purified yeast in 1873, but not until the U.S. Supreme Court decision in *Diamond v. Chakrabarty* in 1980 was a bacterium awarded a patent that was independent of being part of a process. That was soon followed by the patenting of genetically modified crops.

After patents were issued for seeds developed through molecular breeding, the economic relationship between plant breeder and growers was transformed from seed ownership to seed stewardship. Power relations shifted from grower to plant breeder and owner of the intellectual property. This became the grist for opposition to GMOs.

In a 2003 letter to *Nature Biotechnology*, Jerry Cayford of Resources for the Future writes that it is a misleading assumption that the controversy over GMOs is about risk or the extent to which genetically modified crops differ from traditionally bred crops: "What underlies the controversy is whether crop germplasm is [in the] public domain or privately owned through patents on plants and animals. If scientists really want to address the root of opposition to transgenic food, they first need to acknowledge what that underlying root is: monopoly control of the world's food supply."[17] In another letter interchange in *Science*, Cayford states that GMO critics will not "ease their hard line on biosafety" unless patenting is reexamined: "It is the patenting of crops that biotechnology critics find so antidemocratic. To these critics, the patenting of the world's food supply by corporations is an assault on democracy more enormous than any military assault."[18]

The European opposition to patenting animals and plants prompted a thousand farmers to demonstrate before the European Patent Office in Munich on April 15, 2009. The international environmental group Greenpeace challenged a European patent application by Plant Genetic Systems for a transgenic plant that was tolerant to the herbicide Basta. The patent challenge was based on the principle of "ordre public" in article 53(a) of the European Patent Convention that it was immoral to own patents of material that was the common heritage of humankind.[19] The board of appeals of the European Patent Office ruled that the revocation of a patent on environmental grounds under article 53(a) of the 1973 convention required that the hazard be sufficiently substantiated. It would not deny a patent on possible but not conclusively documented risks.[20] Patenting of life forms *sui generis* became the new normal.

Corporate Hegemonic Control

Patents, as a means to gain monopoly control of germplasm, are one of the methods used to achieve hegemony over the agricultural system. The term *biohegemony* has been used by social scientists to describe the methods that biotechnology corporations have used to leverage greater concentration of ownership and control over the food system and to keep agrarian reform of agriculture and agro-ecology movements at bay. The technological control is only one part of the path to hegemony. GMO skeptics are as suspicious of corporate power as they are of the health and safety of the transgenic crops.

For a small group of agricultural seed producers to gain control over a nation's agricultural system, they need to have close political ties to the political elites in the country, be well embedded in international trade organizations, have a strong influence in the agricultural science community and large farmer organizations, and have a strong influential presence in the national media. GMO seed producers need to convince government officials that biotechnology will earn them money in foreign exports and that their support of harmonization, rather than cultural differentiation in biotech policies, will advance their national interests.

One of the strongest examples illustrating biohegemony can be found in Peter Newell's case analysis of biotechnology in Argentina, a country where the agricultural biotechnology sector has amassed hegemonic control. As Newell writes:

> The hegemonic discourse in Argentina regarding agricultural biotechnology is that it represents an important, economically significant, socially beneficial, safe, and environmentally benign technology. This is sustained through government speeches and policy documents, the publicity work of individual companies and associations through seminars, conferences, press conferences, constant advertising in the media aimed at policy and public audiences, and through billboards in the countryside aimed at reaching farmers directly.[21]

For some agricultural communities, adopting GMOs means that they are buying into a new corporate-centric food system where farmers become serfs to the seed producers and their patent-holding companies. In 2004, the residents of Mendocino County, California, voted to ban GMOs from their agricultural land. Measure H—an ordinance titled "Prohibition of the Propagation, Cultivation, Raising, and Growing of Genetically Modified Organisms in Mendocino County"—was directed at those organisms whose "native intrinsic DNA has been intentionally altered or amended with non-species specific DNA." The issues of concern included local control of agriculture (civic agriculture) as opposed to corporate control over local life, equity distribution of power, and contamination of organic farms with GMO pollen. Residents had embraced a food culture that was inconsistent with GMO agriculture.

Risk has not been viewed strictly in terms of the food products but was based on the lack of trust in the GMO enterprise, as has become clear in the STS scholarship: "Measure H proved to be about much more than biotechnology. It served as a lightning rod and symbol of wider social and political issues, many of which reflect tensions between the conventional agricultural system and...civic agriculture...the embedding of local agriculture and food production in the community."[22] Proposals to turn skepticism about GMOs in developing countries to support include a larger role for public research institutions that would forego intellectual property claims for new varieties, similar to what occurred during the Green Revolution.[23]

A case study in *Science Studies* by Pablo Pellegrini of one of the most influential farming organization in Brazil—Movimento Sem Terra (MST)— explores the arguments behind its opposition to GMOs. MST mobilized its anti-GMO actions around issues of uncertainty about health and environment effects, protection of local needs, protection of landraces and agroecology, and concerns over economic concentration, where local farmers lose control to multinational corporations. In its literature, MST states that "GM products and seed research undertaken by corporations only aim to increase their profits and not the population's welfare. The dominance of biotechnology and the use of GMOs are moving towards a world seed oligopoly controlled by just eight major economic groups."[24]

Biohegemony in STS studies also includes control over the sciences. Johan Diels and his colleagues reveal the impact of corporate funding of GMO studies: "In a study involving 94 articles...it was found that the existence of either financial or professional conflict of interest was associated with study outcomes that cast genetically modified products in a favorable light....[A] strong association was found between author affiliation to industry...and study outcomes."[25]

Many factors revealed in the STS scholarship feed the continuing GMO debates. As noted by Renata Motta, "social disputes over GMOs will remain controversial, as there is neither a sole explanation nor a simple solution for them."[26] And neither expertise nor participatory exercises have contributed to a decrease in the controversy.

In conclusion, the STS scholarship has shown that public skepticism or opposition to GMOs is not simply grounded in risk or differences between GMOs and traditional crops. The new technology carries with it new corporate power arrangements, intellectual property controls, unanticipated outcomes, and threats of an increasing concentration and vertical integration of food production. Even without a clear and present danger of GMO risks and with only uncertainties and lack of trusted oversight, the opposition has been built on changes that have intensified a neoliberal political economy of agriculture, reduced the influence of growers and consumers, and centralized power among the oligopoly of molecular breeders and food distributors.

14   Conclusion

This book began with a series of questions about the process and products of molecular breeding, also known as *genetically engineered crops* or *genetically modified organisms* (GMOs). As I noted there, no new technology has ever experienced the type of sustained societal divisiveness that GMOs have experienced. All sides of this debate have a selection bias in what they view as the evidence for their position. A fair-minded science can resolve some (but not all) of the issues. Those who remain skeptical about some claims recognize that science itself may have tacit values or normative presuppositions embedded in the analysis. This is well supported by the science and technology studies (STS) literature. For example, there is no consensus in the scientific definitions for *substantial equivalence*. When a regulatory agency declares a GMO to be substantially equivalent to its parental (non-GE counterpart) crop, it may be embracing trans-scientific concepts masquerading as science.[1] Similarly, the concept of *weight of evidence* is sometimes used as though it is a scientific result. But the weight of evidence in support of a conclusion is rarely ever clearly defined or operationalized.[2]

Many (but not all) of the contested issues raised about genetically modified crops are about risks. Referring back to the initial questions in the introduction, these include the risks of consuming genetically modified crops, lowering productivity, reducing biodiversity, imposing ecological hazards, and thwarting efforts to advance sustainable agriculture.

The study and analysis of risks are largely relegated to technical fields such as risk analysis, decision analysis, and toxicology. These fields require an understanding of potential hazards and of the probability that any single outcome will result from a product or technology. For the moment, let us assume that the assessment of risks is largely or exclusively subject to empirical investigation and logical reasoning. After the results (probability times hazard) are determined, the next question is, What risks are acceptable and at what cost? That part of the risk issue is not scientific. We often have to decide the answer for ourselves as individuals or as a society through a democratic process. I decide whether I shall take a drug with its attendant and often unpredictable side effects in exchange for its potential benefits. Alternatively, a policy sector or a regulatory body will decide whether it will approve a product given its level of risk. If risk assessment were that easy, we could resolve a lot of issues. Risk assessment would be relegated to science, and risk acceptance would be the province of a democratic process and the representatives chosen by that process.

Quite often, however, the evaluation of risk does not fall neatly into the domains of pure science and policy. For one thing, when there is insufficient knowledge, scientists make guesses about the outcome or probability that certain events will occur. These discretionary or transcientific assumptions allow scientists to reach an outcome in the risk assessment. But the assumptions chosen may not be universally accepted or subject to empirical validation. As a result, disagreements in the risk assessments can arise.

In some notable cases, risk assessors have chosen a most-probable case scenario of a toxic event. The public, on the other hand, may choose to make a decision of acceptable risk on a worst-case analysis, where the probabilities are lower but the outcome is more hazardous. Science cannot resolve the framing of the risk problem because it is not a scientific issue.

And we also have to address the risks of not accepting a product or technology. Some prominent voices argue that opposing the development of Golden Rice could result in preventable cases of blindness and mortality. Risks and their acceptance or rejection must be applied fairly in examining pro-GMO and anti-GMO stakeholders. Moreover, the issue of justice demands that we ask, Risks for whom and benefits for whom? From one perspective, some risks to vulnerable developing world farmers and consumers may be less acceptable than the same risks to Western industrial farmers and consumers. Yet some have argued that those dealing with food scarcity should be willing to take more risks in growing and consuming GMOs or in accepting food aid.

There is also the issue of unintended consequences, which are the consequences that cannot be clearly articulated or measured. It is highly problematic to include unintended consequences or acceptable risk in a risk assessment. In the field of technology assessment, the best that can be done is to test the product in sufficiently varied settings at extreme conditions that will reveal any possible unintended consequences (see chapter 9).

When a company sells a food product whose manufacturing process results in unintended consequences for the consumer, the U.S Food and Drug Administration takes action. The agency identifies where the products have been sold, calls for them to be removed from shelves, and requires compensation for anyone harmed. This is possible because all processed food products are labeled and can be traced back to the time and place of manufacture and to their distribution networks. Sometimes raw agricultural products that are contaminated with pathogenic bacteria can be traced to their source. Otherwise, citizens are warned that, until further testing, they should avoid the product. It has been suggested that if a GMO were harmful, it would be discovered by researchers or farmers during the breeding process or by companies prior to commercial distribution. Without labeling and traceability, the unintended effects may be difficult to discover, especially if they are subacute or chronic.

My approach to addressing these questions has been to examine the scientific literature deeply and fairly. I chose not to impose a theoretical perspective on selecting the literature or on following any single line of reasoning. If I was guided by any framework at all, it would be by one of Robert Merton's basic norms of science, "organized skepticism," which influenced my approach to each issue and allowed the evidence to reveal itself from the breadth and depth of the scientific literature.

1. How does traditional plant breeding compare with the plant breeding taking place in biotechnology (molecular breeding)?

Question 1 in the introduction asks about the differences between traditional breeding and molecular breeding. What are the differences and similarities? Is genetic engineering a continuation of the plant breeding that humans have practiced for thousands of years? Or is it qualitatively different and, if so, by what criteria?

It is generally agreed that traditional breeding has limits that can be overcome by molecular breeding. Backcrossing or hybridization would not allow us to get a fish gene and its protein product into a vegetable. Traditional breeding generally rearranges the existing genes within a plant species or taxonomically very close plants. But that oversimplifies the situation. Mutagenesis by radiation or chemicals can create new gene variants or new alleles, but those are still constrained by the existing genotype of the plant. Transplanting new genes from distant species remains a qualitatively new process. As we have seen, there remains uncertainty within the scientific community on the relative risks in these two forms of breeding. European agencies consider molecular breeding to have a higher risk for uncertain effects, and the U.S. agencies consider the risks equivalent and of little concern.

2. What is known about the health assessment of genetically modified crops?

3. What are the arguments that the oversight of bioengineered crops should or should not be stricter than the oversight for traditional crop breeding?

Question 2 in the introduction ask about the health risks of genetically engineered crops. How can we account for differences among scientific studies on animal feeding experiments? Are such studies appropriate for evaluating the health effects of genetically engineered crops? What, if anything, do those experiments reveal about whether GMOs are safe to eat directly or be included in the food chain of processed food? If those experiments are not appropriate, how else are GE crops evaluated? Are there more or greater risks (health or environmental) involved in transferring genes into plants from widely divergent species (such as across genus, families, and even kingdoms) than through intraspecies gene transfer? Question 3 asks whether the novelty of the technology warrants special regulatory oversight. What is the distinction between process-based and product-based regulation? Does it make sense to have special regulations for a process—namely, the use of gene splicing to create GMOs?

There are two ways to approach the answers to these questions. The first is by science, and the second is by public opinion. The National Academies of Sciences, Engineering, and Medicine (NASEM) has consistently written in its reports that there are no unique risks with respect to applying recombinant DNA methods (gene splicing) to developing plant varieties. "No unique risks" does not mean "risk free." NASEM has argued that GMOs should not be treated differently than crops that are traditionally bred. But NASEM's 2004 study of GMO risks states that there are higher risks when genes from distantly related organisms are the source of transplanted DNA. The unanticipated risks are likely to be lower if the foreign genes come from closely related species and much higher if gene splicing is used to transfer genes from distantly related species or if chemical or radiation mutagenesis is used on crop germplasm.

Animal feeding studies have been one of the primary ways that scientists and regulatory bodies have sought to answer the questions about food safety of GMOs. About two dozen published studies have found adverse impacts on animal feeding experiments, whereas hundreds of feeding experiments have not shown adverse effects. The relatively small group of animal experiments that have revealed health effects have been derided by some scientists and organizations for not being a reliable method for determining food safety. The 2004 and 2016 NASEM reports concur that both conventional and molecular forms of breeding have unintended effects and that the unintended effects from GMO breeding and conventional breeding methods are within the same range of probability.

The 2016 NASEM report indicates that there is a lot we do not know about the material differences between a GMO and its parental crop (non-GM counterpart). One way to bridge the gap in knowledge, according to the NASEM study, is to apply omics analysis to the GMOs: "Knowing the variation that occurs naturally in a species one can compare the engineered genome with the reference genome [parental strains] to reveal whether genetic engineering has caused any changes—expected or unintended—and to gain context for assessing whether changes might have adverse effects."[3] Until such an analysis is undertaken, we will not know the answer to the question of whether GMOs are likely to produce greater unintended adverse effects than traditional breeding. Omics analysis can potentially resolve many of the issues of regulation. The 2016 NASEM report proposes a scheme involving four tiers of testing for evaluating transgenic crops that use the full array of omics analysis to compare proteins, metabolites, genomics, and transcriptomics (RNA molecules) between a transgenic and its parental crop.

The report calls for an omics approach for both genetically engineered and conventionally bred crops to determine any unintended changes. The problem is that no regulatory authority requires the omics approach as described in the NASEM report. Moreover, the European regulatory authorities have funded animal feeding studies based on traditional toxi-cological methods. Scientists are beginning to apply the omics analysis in their research, as shown by Yanhua Tan and his colleagues, who studied forty-four unique proteins in the leaves of transgenic and nontransgenic maize. In this study, significant differences were observed, and one of the differentially expressed proteins was identified as a new toxin.[4]

Another omics analysis was completed in 2016 on Roundup-tolerant genetically modified maize. Robin Mesnage and his colleagues performed proteomics and metabolics analyses of GM maize sprayed and unsprayed with Roundup. They found significant differences in the metabolites between the GMO and non-GMO varieties resulting from the transgene: "The transformation process and the resulting expression of a transgenic protein cause a general disturbance in the GM plant and it is clear that NK603 maize is markedly different from its non-GM isogenic line at the proteomic and metabolome levels."[5] The study showed that the GMO and its non-GMO isogenic strain were not substantially equivalent because the chemical enzymatic pathways were different. However, nothing was revealed about the toxicity of the GMO corn in this partial omics analysis.

Omics analysis may reveal changes in the metabolites, proteins, or path-ways of GMOs compared to their non-GMO counterparts, but it does not yield information about toxicology unless it shows increases in an already known toxic product. The question about whether a new transgenic crop will produce novel proteins or metabolites with toxic properties cannot be answered before conducting experiments.

Beyond testing, other considerations in evaluating GMOs are public opinion and consumer trust in the new products. The fact that a majority of the public has supported GMO labeling and that public debate has per-sisted for decades informs us that the public attitude is in favor of greater oversight or at least transparency about GMO-constituted food products.

4. What evidence, if any, is there that genetically modified crops are more productive (produce greater yields) than traditionally bred crops?

Question 4 in the introduction asks about the productivity or yield of GMOs. This question bears on whether GMOs will contribute to a greater supply of food given the same quantity of seeds and other production inputs, such as fertilizer, water, and land.

My analysis reaches the conclusion that the question cannot be answered unambiguously. One must acquire an understanding of the environmental and soil conditions—where the plants are grown and harvested, what the weed and insect resistance is during the time that planting and harvesting take place, and what technologies and inputs are used in an agricultural setting. There is no single or simple answer to the question of GMO yield. Before *Bacillus thuringiensis* (Bt) resistance was discovered in target insects or herbicide resistance in target weeds, there was evidence of higher yields from Bt corn and cotton. Those yields decreased after the resistant weeds and insects spread. In response to GMO pest resistance, new approaches were taken by creating GMOs that genetically engineer more than one pesticide or multiple pathways for herbicide tolerance. One commentator has written that "Blanket conclusions that the technology is a success or failure lack the right level of nuance.... It's an evolving story in India, and we have not yet reached a definitive conclusion."[6] Other commentators have reached a more favorable conclusion: "On average, GM technology adoption has reduced chemical pesticide use by 37%, increased crop yields by 22%, and increased farmer profits by 68%. Yield gains and pesticide reductions are larger for insect-resistant crops than for herbicide-tolerant crops. Yield and profit gains are higher in developing countries than in developed countries."[7]

The scientific literature about GMO yields provides conflicting results, depending on who is reporting the studies, from which countries they are reporting the yields, and what sources and methods are used in compiling the data. In one study, the authors undertook a meta-analysis of 147 original studies that examined the most important genetically engineered crops, including herbicide-tolerant soybeans, maize, and cotton and insect-resistant maize and cotton. The study reported that, on average, GE crop yields increased from non-GE yields by 21 percent. These yield increases were more pronounced for the developing world than the developed world.[8]

In contrast, another group of scientists obtained yield data of maize, rapeseed, soybean, and cotton from the United Nations Food and Agriculture Organization (FAO) database for the United States, Canada, and Western Europe for 1961 to 2010. They reported that the United States had only marginally significant higher yields than Europe, despite the former's much greater dependence on GE crops: "in recent years W. Europe has had similar and even slightly higher yields than the United States despite the latter's use of GM varieties.... Notwithstanding claims to the contrary ... there is no evidence that GM biotechnology is superior to other biotechnologies in its potential to supply calories."[9] From their literature review, Alan B. Bennett, Cecilia Chi-Ham, Geoffrey Barrows, Steven Sexton, and David Zilberman found that, on average, GE cotton and GE maize increased yield, reduced pest damage, reduced pesticide use, and increased farm profitability. GE soybeans, however, showed declines, and many farms were outliers and did not show the yield increases in GE cotton and GE maize.[10]

There is no clear consensus that GMOs are inherently better or worse than non-GMO varieties in terms of yield. Comparing different growth regions and practices will not provide a useful answer to a complex question. The literature suggests that each region, practice, climate, pest density, pesticide use, farm management practice, and transgenic seed will contribute to delivering a yield for that GM seed. Controlled greenhouse studies of GMOs and non-GMOs, everything else being equal, may provide useful information. But that information may not transfer to actual growing conditions in the field.

5. What distinctions are there, if any, between the environmental impacts of GMOs and traditionally bred crops?

Question 5 in the introduction asks whether GMOs are more, less, or neutrally favorable to the environment. Are some GMOs more favorable environmentally than traditional crops? Do GMOs have any unique impacts on biodiversity, beyond the impacts of traditional crops? Will GMOs contribute or become an obstacle to sustainable agriculture? Will they reduce or increase chemical pesticide use? Will they enhance, be neutral to, or reduce biodiversity? Will they support healthy and sustainable soil? Can this be predicted? Can it be ascertained from specific cases? What does the science tell us? Where is there consensus within the science?

One of the popular claims made about insect-resistant crops is that they will reduce the use of chemical pesticides. It is a plausible scenario. If insects are deterred or killed by eating a transgenic crop with a toxin gene, chemical pesticides will not need to be sprayed on the crop. This is similar to the idea that some plant lectins protect plants from certain insects. This assumes that insects will not become resistant to the Cry toxins in Bt crops. What do we know about the reduction of chemical pesticides on a farm scale where insect-resistant transgenic crops have been applied?

Charles M. Benbrook undertook a systematic study of pesticide use with genetically modified crops.[11] We discuss his conclusions about the use of herbicides in conjunction with herbicide-tolerant crops in chapter 5. Benbrook found that Bt-transgenic corn and cotton displaced about 123 million pounds of chemical insecticides from 1996 to 2011. He also notes that every plant in a Bt corn or cotton field is producing insecticidal proteins within its cells. Benbrook calculated the amount of Bt protein toxins that are produced within all the plant components in the Bt corn and cotton crops in the United States. He estimates that the total Cry protein production in U.S. agriculture is about 3.7 pounds per acre, which amounts to nineteen times greater than the average chemical insecticide application in 2010. There is no consensus among entomologists, notes Benbrook, on whether Bt toxins produced in transgenic plants should be counted in determining whether Bt crops have reduced overall insecticide use. No disagreement can be found that insect-resistant transgenic crops have had an initial effect on reducing chemical pesticides. The amount of reduction will depend on the degree to which insect resistance to Bt develops. Finally, Benbrook argues that if Bt crop technology becomes more sophisticated so that the Bt endotoxins appear only in those plant tissues that are actively attacked by insects and only at the times that insects are feeding, then Bt crops could be compatible with integrated pest management. Thus far, biotechnology has not reached this level of sophistication.

A growing concern over insect resistance to Bt crops has led some farmers to return to chemical insecticides. Bruce E. Tabashnik, Thierry Brévault, and Yves Carrière review the effect of Bt crops on insect resistance based on monitoring data covering two decades from twenty-four cases in eight countries, after a billion acres had been planted. They conclude that "the number of major target pests with some populations resistant to Bt crops and reduced efficacy reported surged from one to five."[12]

Edward D. Perry, Federico Ciliberto, David Hennessy, and GianCarlo Moschini sum up the answer to whether pesticide use has increased or decreased in GE crops: "For both soybean and maize GT [glyphosate-tolerant] adopters use increasingly more herbicides relative to nonadopters, whereas adopters of IR [insect-resistant] maize use increasingly less herbicides."[13]

What about GMOs and sustainability? Both the pro-GMO and anti-GMO communities have sought ownership of the concept of sustainability as a framework for advancing their advocacy. It is reminiscent of the biotechnology debates in the 1980s, when similar struggles took place over ownership of the meaning of *natural food*. In the contemporary world, the term *sustainability* implies actions taken to protect the biosphere from despoliation and declines in the biodiversity of all life. Sustainable agriculture seeks to protect and preserve soil and water quality as well as the ecosystems within which farms are embedded. It also has been associated with low-chemical-input agriculture. A. Wendy Russell maintains that "sustainability is a highly contested term, in general, and in the GM debate."[14] It also is recognized that sustainability is not a scientific concept but a societal ideal and one for which there is no clear definition. Douglas H. Constance argues that the ambiguity of the concept of sustainability allows seed oligopolies to turn it to their advantage: "because the concept of sustainability is deeply contested, agribusiness is able to exploit the ambiguity surrounding the definition of *sustainable* and exercise power in attempts to frame sustainable agriculture in their favor."[15]

Critics of genetically engineered crops are adamant that GE seeds are incompatible with sustainable agriculture. The seeds are produced, patented, and distributed by multinational corporations that support a neoliberal, high-input model of agricultural production. Advocates of GMOs use sustainability criteria for each product and compare the GMO with its parental strain (its non-GM counterpart).

Sustainability is not the outcome of a utopian world. It is one step in the right direction. We can meaningfully ask whether practice A or product B is more sustainable than practice A' or product B'. There are no absolute measures, only comparative ones. Russell highlights the point: "it is not feasible to ask whether a particular system, industry or technology is 'sustainable' or 'unsustainable,' but useful to consider whether it is associated with a tendency towards or away from sustainability."[16]

It is not be possible to answer the question of whether conventional breeding is more sustainable than molecular breeding without considering what the two methods produce, what they substitute for, and what agricultural system they fit into. If we focus only on the process and examine it according to the four factors associated with sustainability cited by the 2016 NASEM report (health, environment, social, and economic), the task would be formidable. The process would have to be evaluated by sustainability criteria as an occupational, environmental, and social activity. Most discussions of sustainability are about the products and the systems they support. The reductionist approach asks whether Bt corn, raised without chemical pesticides, is more or less sustainable than traditionally bred corn raised with pesticides. If that were the only comparison (where soil, water resources, and humans are saved from pesticide exposure), the answer could be in the affirmative. Because organic farming does not permit the use of genetically engineered crops, the question of whether GE seeds could make organic agriculture more sustainable is not meaningful.

Echoing the 2016 NASEM report, a complete analysis of sustainability would have to cover all four of the factors. We would have to know whether the Bt crops were more or less dangerous to nonintended species or whether they provided added social and economic benefits to farmers, especially small farmers. The NASEM report dismisses any rhetorical claims made about GMO sustainability in its findings: "Although emerging genetic-engineering techniques have the potential to assist in achieving a sustainable food system, broad and rigorous analyses will be necessary to determine the long-term health, environment, social and economic outcomes of adding specific crops and traits to an agroecosystem."[17]

6. Have any commercialized GMO crops been designed to improve a crop's nutritional quality, flavor, or other attributes valued by consumers or public health advocates (such as through biofortification)?

Question 6 in the introduction, regarding nutritional improvements in GMO crops, is addressed in part in chapter 12 on Golden Rice, which describes the efforts that were made to bring to the market the first biofortified staple crop designed to save lives and eyesight in vitamin A–deprived countries. After nearly two decades, Golden Rice has not been approved for commercial markets. Nevertheless, the scientific literature shows evidence of its great promise. As an example, one 2015 review in the *International Journal of Molecular Sciences* states: "Great progress has been made over the past decade with respect to the application of biotechnology to generate nutritionally improved food crops. Biofortified staple crops such as rice, maize and wheat harboring essential micronutrients to benefit the world's poor are under development as well as new varieties of crops which have the ability to combat chronic disease."[18]

Notwithstanding such expressions of enthusiasm, biofortified GMOs have yet to be proven a public health success. Beyond the science, acceptance will involve unambiguous health assessments, philanthropic support, adaptation to local cultivars, and cultural adoption.

7. What are the critical issues regarding the demands for and against mandatory labeling of GMOs?

Question 7 in the introduction asks about the labeling of GMOs, which is covered in chapter 10. Is there a rational basis for labeling? How does the European Union compare with the United States in regard to labeling genetically engineered crops and GMO food? What federal or state initiatives have been taken to label GMO crops in the consumer market?

Rationality in this case depends on what the starting assumptions are. Both the European Union and the United States have reached decisions in favor of labeling in their unique ways. Their respective decisions were not crafted in rational arguments but rather based on the power of public opinion and lobbying. The question of whether the process of molecular breeding is likely to produce more unanticipated risks than traditional breeding will have to await future study as more rigorous analyses of the products of both types of breeding are compared. Sometimes it takes years to understand the impact of new food technologies and their products. This has certainly been the case with fast foods, processed food, sugary drinks, and transfats.

Public opinion polls favor labeling by up to 90 percent. Most food manufacturers are against it. Congress passed what might be called a "stealth labeling law" without penalties if it is violated. Either a smartphone-readable code or phone number can be used by the consumer to decipher whether a product has been grown or made with genetically modified products. Nothing on the packaging will be as apparent as the salt or calorie content of food. The federal labeling law preempts any state labeling legislation. Although the federal labeling mandate solves the problem of a patchwork of state labeling laws, until it is fully implemented, its value to consumers will not be known.

8. Can GMO and non-GMO agriculture coexist?

Finally, a new question addresses the cohabitation of GMO and non-GMO agriculture. Organic farms are certified as non-GMO, and organic farmers have raised concerns that GMO pollen wafting onto their farms can threaten their organic certification. Miguel A. Altieri considers the coexistence of GMOs and non-GMOs to be a myth: "The first important argument against the concept of coexistence is that the movement of transgenes beyond their intended destinations and hybridization with weedy relatives and contamination of other non-GM crops is a virtual certainty."[19] A 2010 headline in *Nature News* provides grist for Altieri's concerns: "GM Crop Escapes into the American Wild."[20] Researchers reported at an Ecological Society of America conference that transgenic canola was growing freely in parts of North Dakota for the first time.

There are no systematic studies of GMO contamination of organic farms in the United States, only anecdotal reports and a U.S. Department of Agriculture survey. In February 2006, the Economic Research Service (ERS) of the USDA issued a report on the coexistence of GMOs with non-GMO crops. In 2008 and 2014, the ERS carried out surveys of certified organic corn producers. From the 2008 survey, it learned that 18 percent of the organic corn farmers in the United States had their corn production tested for transgenic components. According to the survey, 1 percent of the certified organic farmers reported that their food-grade corn was rejected, and 2 percent reported that their feed-grade corn was rejected by a buyer as a result of GMO contamination in 2010 or earlier.

In the 2014 National Organic Survey, farmers were asked if they had "experienced economic losses that you can document due to unintended presence of GMO material in an organic crop that you have produced for sale."[21] In this survey, 1 percent of the U.S. certified organic farmers representing eighty-seven organic farms in twenty states declared economic losses from the presence of GMO material in their crops from 2011 to 2014. The total estimate of the losses was around $6 million. Some states had losses of less than 1 percent in their organic farms, and others reported losses of 6 to 7 percent in their organic farms. The largest number of farmers reporting losses lived in Illinois, where sixteen farms had losses that averaged $38,884 per farm.

Organic farms remain a small, albeit growing part of U.S. agricultural production. In 2012, 390 million cropland acres were under cultivation in the United States, of which 182 million acres were planted with genetically engineered seed, 90 percent in corn and soybean. In contrast, 5.4 million acres were classified as certified organic farms in 2011, and only 0.3 percent (234,000 acres) of U.S. corn and 0.2 percent (132,000 acres) of U.S. soybean were organic.

Thus far, the commingling of GMOs and non-GMOs has cost eighty-seven farms about $6 million: "Nationwide, a total of 87 farms in 20 States reported an economic loss in at least 1 year between 2011 and 2014. These farmers reported a total of nearly $6.1 million in economic losses during this period, accounting for an estimated 0.4 percent of the total value of farm sales for all 9 crops with GE counterparts during 2011–2014. The average economic loss from unintended GE presence in organic crops varied substantially by State."[22] Organic famers also paid an unknown amount of litigation costs when Monsanto discovered GMO plants on their cropland.[23]

The United States has no national policy covering the coexistence of GMO and non-GMO farms. The situation is quite different in Europe, where the European Union has requested all member states to adopt measures to prevent the admixture of GMO and non-GMO crops. Europe has a labeling requirement for GMOs, where the standard admixture for most countries cannot be more than 0.9 percent GMOs in non-GMO crops.[24]

The European Commission defines the separation of GMO and non-GMO crops as follows: "Co-existence refers to the ability of farmers to make a practical choice between conventional, organic and GM-crop production, in compliance with the legal obligations for labelling and/or purity standards."[25] The Commission held that member states may take appropriate measures to prevent the commingling of commercial GMO products with non-GMO products. The coexistence of GMO and non-GMO agriculture is an established policy in many EU countries. The Netherlands, for example, established the Dutch Coexistence Committee to set standards for separations of GMO and non-GMO crops based on outcrossing studies. In addition to establishing buffer zones that are large (or extensive) enough to prevent outcrossing, other biotechnological methods are used to prevent horizontal gene flow from GMOs to non-GMOs.[26]

A decade ago, there was controversy among scientists about whether it was possible to prevent outcrossing from GMO crops to non-GMO crops. Clemens C. M. Van De Wiel and L. A. P. Lotz, in a study funded by the Dutch Ministry of Agriculture, find that the measures to sustain coexistence at a 0.9 percent EU labeling threshold between GM with non-GM agriculture with sugar beet, potato, and maize were plausible with a 25 meter separation, whereas there was skepticism about coexistence with oilseed rape due to volunteers and wild relatives.[27] Alessandro Chiarabolli describes the Portuguese regulations and claims that they have implemented the most complete system of coexistence, which has not experienced any difficulties in segregating harvests of GMOs and non-GMOs.[28]

The Europeans are invested in protecting their organic farms while also providing opportunities for the commercialization of several GMO varieties. Some countries, like Portugal, have established compensation funds to cover damages caused by the accidental contamination of non-GMO crops. In the United States, organic and other non-GMO farms are largely on their own.

Finally, in 2003, Marion Nestle wrote that "Overall, the role of genetically modified foods in these larger aspects of the food system is as yet uncertain and unlikely to be known for some time to come."[29] Fourteen years later, over 90 percent of all the soy, corn, and cotton grown in the United States is genetically modified. Other approved genetically engineered crops include sugar beets, alfalfa, canola, papaya, summer squash, potato, and apples. This suggests that the U.S. approval process is far from uncertain, notwithstanding the more cautious approval in Europe, even while many scientific questions are still unanswered. GMOs remain one of the most persistent and resilient technological controversies in modern history.

Notes

Introduction

1. Francis Bacon, "The New Atlantis," in *Sir Thomas More, The Utopia ... 1551, Francis, Lord Bacon, The New Atlantis, 1622*, ed. H. Goitein (London: Routledge [1925]).

2. Sheldon Krimsky, *Biotechnics and Society* (New York: Praeger, 1991), 193.

3. U.S. Food and Drug Administration, "Statement of Policy: Foods Derived from New Plant Varieties," *Federal Register* 57 (May 29, 1992): 22,984.

4. U.S. Food and Drug Administration, "Premarket Notice Concerning Bioengineered Foods: Proposed Rule," *Federal Register* 66 (January 18, 2001): 4706–4738.

5. U.S. Food and Drug Administration, "Premarket Notice."

6. Jon Entine, "The Debate about GMO Safety Is Over, Thanks to a New Trillion-Meal Study," *Forbes*, September 17, 2014, https://www.forbes.com/sites/jonentine/2014/09/17/the-debate-about-gmo-safety-is-over-thanks-to-a-new-trillion-meal-study/#6250f0be8a63.

7. Angelika Hilbeck, Rosa Binimelis, Nicolas Defarge, et al., "No Scientific Consensus on GMO Safety," *Environmental Sciences Europe* 27, no. 4 (2015): 1–6.

8. National Academies of Sciences, Engineering, and Medicine, *Genetically Engineered Crops: Experiences and Prospects* (Washington, DC: National Academy Press, 2016).

9. The search was carried out on August 27, 2016.

Chapter 1: Traditional Plant Breeding

1. Catherine Preece, Alexandra Livarda, Pascal-Antoine Christin, Michael Wallace, Gemma Martin, Michael Charles, Glynis Jones, Mark Rees, and Colin P. Osborne, "How Did the Domestication of Fertile Crescent Grain Crops Increase Their Yields?," *Functional Ecology* 31 (2017): 387–397.

2. George Acquaah, *Principles of Plant Genetics and Breeding*, 2nd ed. (New York: Wiley-Blackwell, 2012), chap. 2.

3. John Bingham, "The Achievement of Conventional Plant Breeding," *Philosophical Transactions of the Royal Society London B* 292 (1981): 441.

4. Hakan Ulukan, The Evolution of Cultivated Plant Species: Classical Plant Breeding versus Genetic Engineering," *Plant System Evolution* 280 (2009): 123.

5. Acquaah, *Principles of Plant Genetics*, chap. 2.

6. Tania Carolina Camacho Villa, Nigel Maxtel, Maria Scholten, and Brian Ford-Lloyd, "Defining and Identifying Crop Landraces," *Plant Genetic Resources: Characterization and Utilization* 3, no. 3 (2005): 373–384.

7. Lawrence Busch, William B. Lacy, Jeffrey Burkhardt, and Laura B. Lacy, *Plants, Power, and Profit: Social, Economic, and Ethical Consequences of the New Biotechnologies* (Cambridge, MA: Blackwell, 1991), 58.

8. National Research Council, *Genetic Engineering of Plants: Agricultural Research Opportunities and Policy Concerns* (Washington, DC: National Academy Press, 1984), 5.

9. National Research Council, *Safety of Genetically Engineered Foods: Approaches to Assessing Unintended Health Effects* (Washington, DC: National Academy Press, 2004), 24.

10. Ulukan, "The Evolution of Cultivated Plant Species," 135.

11. National Research Council, *Genetic Engineering of Plants*, 5.

12. Herbert F. Roberts, "The Founders of the Art of Breeding—II," *Journal of Heredity* 10, no. 4 (1919): 229–239.

13. Noel Kingsbury, *Hybrid: The History of Plant Breeding* (Chicago: University of Chicago Press, 2009).

14. Kingsbury, *Hybrid*, 257.

15. H. C. Sharma, "How Wide Can a Wide Cross Be?," *Euphytica* 82 (1995): 43.

16. National Research Council, *Safety of Genetically Engineered Foods*, 26.

17. Sharma, "How Wide Can a Wide Cross Be?," 59.

18. Sharma, 59.

19. Kingsbury, *Hybrid*, 257.

20. Sharma, "How Wide Can a Wide Cross Be?," 56.

21. Rodomiro Ortiz and Dirk Vuylsteke, "Recent Advances in Musa Genetics, Breeding and Biotechnology," *Plant Breeding Abstracts* 66, no. 10 (1996): 1355–1363.

22. Stephen P. Moose and Rita H. Mumm, "Molecular Plant Breeding as the Foundation for the Twenty-first-Century Crop Improvement," *Plant Physiology* 147 (2008): 969–977.

23. Michael Hansen, Lawrence Busch, Jeffrey Burkhardt, William B. Lacy, and Laura R. Lacy, "Plant Breeding and Biotechnology," *BioScience* 36, no. 1 (1986): 29–39.

Chapter 2: Molecular Breeding

1. I use the term *molecular breeding* to mean the application of molecular biology tools in plant and animal breeding.

2. Yan Zhang, Thomas Lubberstedt, and Mingliang Xu, "The Genetic and Molecular Basis of Plant Resistance to Pathogens," *Journal of Genetics and Genomics* 40, no. 1 (2013): 23–35.

3. M. D. Chilton, M. H. Drummond, D. J. Merlo, D. Sciaky, A. L. Montoya, M. P. Gordon, and E. W. Nester, "Stable Incorporation of Plasmid DNA into Higher Plant Cells: The Molecular Basis of Crown Gall Tumorigenesis," *Cell* 11, no. 2 (1985): 263–271.

4. M. Haas, M. Bureau, A. Geldreich, and P. Keller, "Cauliflower Mosaic Virus: Still in the News," *Molecular Plant Pathology* 3, no. 6 (2002): 419–429.

5. Nancy Podevin and Patrick du Jardin, "Possible Consequences of the Overlap between the CAMV35S Promoter Regions in Plant Transformation Vectors Used and the Viral Gene VI in Transgenic Plants," *GM Crops and Food* 3, no. 4 (2012): 296–300.

6. John C. Sanford, Theodor M. Klein, Edward D. Wolf, and Nelson Allen, "Delivery of Substances into Cells and Tissues: A Projectile Bombardment Process," *Particulate Science Technology* 5, no. 1 (2007): 27–37.

7. Behrooz Darbani, Safar Farjnia, Mahmoud Toorchi, Saeed Zakerbostanabad, Shahin Noeparvar, and C. Neal Stewart Jr., "DNA-Delivery Methods to Produce Transgenic Crops," *Biotechnology* 7, no. 3 (2008): 385–402.

8. Jonathan R. Latham, Allison K. Wilson, and Ricarda A. Steinbrecher, "The Mutational Consequences of Plant Transformation," *Journal of Biomedicine and Biotechnology* 2006 (2006): 1–7.

9. Jan G. Schaart, Clemens C. M. van de Wiel, Lambertus A. P. Lotz, and Marinus J. M. Smulders, "Opportunities for Products of New Plant Breeding Techniques," *Trends in Plant Science* 21, no. 5 (2016): 438.

10. Samriti Sharma, Rajinder Kaur, and Anupama Sing, "Recent Advances in CRISPR/Cas Medicated Genome Editing for Crop Improvement," *Plant Biotechnology Reports* 11 (2017): 193–207.

11. Leena Arora and Alka Narula, "Gene Editing and Crop Improvement Using CRISPR-Cas9 System," *Frontiers in Plant Science* 8, no. 1932 (2017): 1–21.

12. Schaart et al., "Opportunities for Products."

13. Noel J. Sauer, Jerry Mozoruk, Ryan B. Miller, Zachary J. Warburg, Keith A. Walker, Peter R. Beetham, Christian R. Schöpke, and Greg F. W. Gocal, "Oligonucleotide-Directed Mutagenesis for Precision Gene Editing," *Plant Biotechnology Journal* 14 (2016): 496–502.

14. Hongwei Hou, Neslihan Atlihan, and Zhen-Xiang Lu, "New Biotechnology Enhances the Application of Cisgenesis in Plant Breeding," *Plant Genetics and Genomics* 5, no. 389 (2014): 1–5.

15. Kaare M. Nelsen, "Transgenic Organisms: Time for Conceptual Diversification," *Nature Biotechnology* 21 (2003): 227–228.

16. Nelsen, "Transgenic Organisms," 228.

17. Felix Wolter and Holga Puchta, "Knocking Out Consumer Concerns and Regulator's Rules: Efficient Use of CRISPR/Cas Ribonucleoprotein Complexes for Genome Editing in Cereals," *Genome Biology* 18, no 1 (2017): 43–45.

Chapter 3: Differences between Traditional and Molecular Breeding

1. Sheila Jasanoff, "Product, Process, or Program: Three Cultures and the Regulation of Biotechnology," in *Resistance to New Technology: Nuclear Power, Information Technology and Biotechnology*, ed. Martine Bauer, 311–331 (Cambridge, UK: Cambridge University Press, 1995).

2. Giovanni Tagliabye, "Product, Not Process! Explaining a Basic Concept in Agricultural Biotechnologies and Food Safety," *Life Science, Society and Policy* 13, no. 3 (2017): 1–9.

3. Douglas A. Kysar, "Preference for Processes: The Process/Product Distinction and the Regulation of Consumer Choice," *Harvard Law Review* 118, no. 2 (2004): 558.

4. National Academy of Sciences, *Research with Recombinant DNA: An Academy Forum, March 7–9, 1977* (Washington, DC: National Academy of Sciences, 1977).

5. National Research Council, *Introduction of Recombinant DNA-Engineered Organisms into the Environment: Key Issues* (Washington, DC: National Academy Press, 1987), 23.

6. National Academies of Sciences, Engineering, and Medicine (NASEM), "Report in Brief: Genetically Engineered Crops: Experiences and Prospects," May 2016, 1, https://nas-sites.org/ge-crops/2016/05/16/report-in-brief.

7. National Research Council, *Introduction of Recombinant DNA-Engineered Organisms*, 17.

8. Hakan Ulukan, "The Evolution of Cultivated Plant Species: Classical Plant Breeding versus Genetic Engineering," *Plant System Evolution* 280 (2009): 133–142.

9. Stephen P. Moose and Rita H. Mumm, "Molecular Plant Breeding as the Foundation for Twenty-first Century Crop Management," *Plant Physiology* 147 (2008): 969.

10. Michael K. Hansen, "Genetic Engineering Is Not an Extension of Conventional Plant Breeding," Consumer Policy Institute, Consumers Union, January 2000, https://consumersunion.org/wp-content/uploads/2013/02/wide-crosses.pdf.

11. Nina Federoff and Nancy Marie Brown, *Mendel in the Kitchen: A Scientist's View of Genetically Modified Foods* (Washington, DC: Joseph Henry Press, 2004), 127.

12. Hansen, "Genetic Engineering," 3.

13. Joy Bergelson, Colin B. Purrington, and Gale Wichmann, "Promiscuity in Transgenic Plants," *Nature* 395 (1998): 25; Tomoko Inose and Kousaku Murata, "Enhanced Accumulation of Toxic Compound in Yeast Cells Having High Glycolytic Activity: A Case Study on the Safety of Genetically Engineered Yeast," *International Journal of Food Science and Technology* 30 (1995): 141–146.

14. Federoff and Brown, *Mendel in the Kitchen*, citing the National Academy of Sciences, 150.

15. Maria Montero, Anna Coll, Anna Nadal, Joaquima Messeguer, and Maria Pia, "Only Half the Transcriptomic Differences between Resistant Genetically Modified and Conventional Rice Are Associated with the Transgene," *Plant Biotechnology Journal* 9 (2011): 693–702.

16. Hakan Ulukan, "The Evolution of Cultivated Plant Species: Classical Plant Breeding Versus Genetic Engineering," *Plant Syst Evol* 280 (2009):133–142.

17. Marcin Filipecki and Stefan Malepszy, "Unintended Consequences of Plant Transformation: Molecular Insight," *Journal of Applied Genetics* 47, no. 4 (2006): 277–286.

18. John Fagan, Michael Antoniou, and Clare Robinson, "What Is Genetic Engineering? An Introduction to the Science," in *The GMO Deception*, ed. S. Krimsky and J. Gruber (New York: Skyhorse Press, 2014), xxxiv.

19. Filipecki and Malepszy, "Unintended Consequences."

20. Hansen, "Genetic Engineering," 7.

21. Hongwei Hou, Neslihan Atlihan, and Zhen-Xiang Lu, "New Biotechnology Enhances the Application of Cisgenesis in Plant Breeding," *Frontiers in Plant Science* 5, no. 389 (2014): 1–6.

22. Hou, Atlihan, and Lu, "New Biotechnology," 1.

23. Hou, Atlihan, and Lu, 3.

24. Sheldon Krimsky, "A Worst Case Experiment," chap. 18 in *Genetic Alchemy: The Social History of the Recombinant DNA Controversy* (Cambridge, MA: MIT Press, 1982).

25. Editorial, "Crop Conundrum: The EU Should Decide Definitively Whether Gene-Edited Plants Are Covered by GM Laws," *Nature* 528 (December 17, 2015): 307–308.

26. Sheldon Krimsky and Alonzo Plough, *Environmental Hazards: Communicating Risks as a Social Process* (Dover, MA: Auburn House, 1988), 306.

27. Hannes R. Stephan, *Cultural Politics and the Transatlantic Divide over GMOs* (New York: Palgrave Macmillan, 2015).

Chapter 4: Early Products in Agricultural Biotechnology

1. Gregory A. Tucker and Donald E. Grierson, "Synthesis of Polygalacturonase during Tomato Fruit Ripening," *Planta* 155 (1982): 64–67.

2. Each gene in the plant cell consists of two strands (a double helix)—the sense strand and the antisense strand. Only one strand is used to make mRNA. If the cell has two DNA coding sequences, one the mirror image of the other, then the cell will make an mRNA molecule, ordinarily single stranded with a double-stranded segment. When part of the mRNA molecule forms a duplex with some antisense mRNA (where the sense mRNA is complementary to and bonds with the antisense mRNA), the plant's cell machinery is unable to recognize the sense mRNA molecule (it sees it as a foreign molecule) and thus prevents the translation process to synthesize the PG enzyme. Stuti Giupta, Ravindra Pal Singh, Nirav Rabadia, Gaurang Patel, and Hiten Panchal, "Antisense Technology," *International Journal of Pharmaceutical Sciences Review and Research* 9, no. 2 (2011): 38–45.

3. Animal and Plant Health Inspection Service, U.S. Department of Agriculture, "Addition of Two Genetically Engineered Tomato Lines to Determination of Non-regulated Status for Calgene, Inc.," *Federal Register* 60, no. 145 (July 28, 1995): 38, 788–38,789.

4. Belinda Martineau, *First Fruit: The Creation of the Flavr Savr Tomato and the Birth of Biotech Food* (New York: McGraw-Hill, 2001), 113.

5. Matthew Kramer, Rick Sanders, Hassan Bolkan, Curtis Waters, Raymond E. Sheeny, and William R. Hiatt, "Postharvest Evaluation of Transgenic Tomatoes with Reduced Levels of Polygalacturonase: Processing, Firmness and Disease Resistance," *Postharvest Biology and Technology* 1, no. 3 (1992): 241–255; Matthew G. Kramer and Keith Redenbaugh, "Commercialization of a Tomato with an Antisense Polyglacturonase Gene: The Flavr Savr™ Tomato Story," *Euphytica* 79 (1994): 293–297.

6. Nina Federoff and Nancy Marie Brown, *Mendel in the Kitchen: A Scientist's View of Genetically Modified Foods* (Washington, DC: Joseph Henry Press, 2004), 175.

7. Kramer and Redenbaugh, "Commercialization of a Tomato with an Antisense Poly-galacturonase Gene."

8. Martineau, *First Fruit*, 187.

9. Martineau, 4.

10. Belinda Martineau, "Food Fight," *The Sciences* 41, no. 2 (2001): 28.

11. Steven E. Lindow, "Methods Preventing Frost Injury Caused by Epiphytic Ice-Nucleation-Active Bacteria," *Plant Disease* 67, no. 3 (1983): 327–333.

12. Leroy R. Maki, Elizabeth L. Galyan, Mei-Mon Chang-Chien, and Daniel R. Caldwell, "Ice Nucleation Induced by *Pseudomonas syringae*," *Applied Microbiology* 28, no. 3 (1974): 456–459.

13. R. M. Skirvin, E. Kohler, H. Steiner, D. Ayers, A. Laughman, M. A. Norton, and M. Warmund, "The Use of Genetically Engineered Bacteria to Control Frost on Strawberries and Potatoes: Whatever Happened to All That Research?," *Scientia Horticulturae* 84 (2000): 181.

14. Office of Technology Assessment, U.S. Congress, *New Developments in Biotechnology: Field-Testing Engineered Organisms* (Washington, DC: U.S. Government Printing Office, May 1988), 95.

15. Skirvin et al., "The Use of Genetically Engineered Bacteria to Control Frost on Strawberries and Potatoes," 187.

16. Skirvin et al., 184.

17. John Love and William Lesser, "The Potential Impact of Ice-Minus Bacteria as a Frost Protectant in New York Fruit Production," *Northeastern Journal of Agricultural and Resource Economics* 18, no. 1 (1989): 26–34.

Chapter 5: Herbicide Tolerant Transgenic Crops

1. Luca Lombardo, Gerardo Coppola, and Samanta Zelasco, "New Technologies for Insect-Resistant and Herbicide Tolerant Plants," *Trends in Biotechnology* 34, no. 1 (2016): 49–57.

2. Charles A. Hamner and Harold B. Tukey, "The Herbicidal Action of 2,4 Dischloro-phenoxyacetic and 2,4,5 Trichlorophenoxy Acetic Acid on Bindweed," *Science* 100, no. 2590 (1944): 154–155.

3. Rachel Carson, *Silent Spring* (New York: Houghton Mifflin, 1962, 1994).

4. Barbara J. Mazor and S. Carl Falco, "The Development of Herbicide Resistant Crops," *Plant Physiology and Plant Molecular Biology: Annual Reviews* 40 (1989): 443.

5. Martin Paul Krayer von Krauss, Elizabeth A. Casman, and Mitchell J. Small, "Elicitation of Expert Judgments of Uncertainty in the Risk Assessment of Herbicide-Tolerant Oilseed Crops," *Risk Analysis* 24, no. 6 (2004): 1523.

6. Charles M. Benbrook, "Trends in Glyphosate Herbicide Use in the United States and Globally," *Environmental Science Europe* 28, no. 3 (2016): 1–15.

7. Jorge Fernandez-Cornejo, Seth Wechsler, Mike Livingston, and Lorraine Mitchell, "Genetically Engineered Crops in the United States," Economic Research Report No. 162, Economic Research Service, U.S. Department of Agriculture, February 2014, p. 6, www.ers.usda.gov/media/1282246/err162.pdf.

8. Benbrook, "Trends in Glyphosate Herbicide Use in the United States and Globally."

9. Ganesh M. Kishore, Stephen R. Padgette, and Robert T. Fraley, "History of Herbicide-Tolerant Crops, Methods of Development and Current State of the Art: Emphasis on Glyphosate Tolerance," *Weed Technology* 6 (1992): 628.

10. Kishore, Padgette, and Fraley, "History of Herbicide-Tolerant Crops."

11. Fernandez-Cornejo et al., "Genetically Engineered Crops in the United States," 16.

12. Doug Gurian-Sherman, "Failure to Yield: Evaluating the Performance of Genetically Engineered Crops," Union of Concerned Scientists, Cambridge, MA, April 2009, 17.

13. Roger W. Elmore, Fred W. Roeth, Lenis A. Nelson, Charles A. Shapiro, and Robert N. Klein, "Glyphosate-Resistant Soybean Cultivar Yields Compared with Sister Lines," *Agronomy Journal* 93 (2001): 408–412.

14. Fernandez-Cornejo et al., "Genetically Engineered Crops in the United States," 22.

15. Fernandez-Cornejo et al., 6.

16. Fernandez-Cornejo et al., 24.

17. Fernandez-Cornejo et al., 24.

18. Environmental Protection Agency, "Registration of Enlist Duo," https://www.epa.gov/ingredients-used-pesticide-products/registration-enlist-duo.

19. Charles M. Benbrook, "Impact of Genetically Engineered Crops on Pesticide Use in the U.S.: The First Sixteen Years," *Environmental Sciences Europe* 24 (2016): 7.

20. Edward D. Perry, Federico Ciliberto, David A, Hennessy, and GianCarlo Moschini, "Genetically Engineered Crops and Pesticide Use in U.S. Maize and Soybeans," *Science Advances* 2, no. 8 (2016).

21. Benbrook, "Impact of Genetically Engineered Crops on Pesticide Use in the U.S.," 1.

22. Sheldon Krimsky and Roger Wrubel, *Agricultural Biotechnology and the Environment* (Chicago: University of Illinois Press, 1996), 47.

23. Jerry M. Green, "The Benefits of Herbicide Resistant Crops," *Pest Management Science* 68 (2012): 1323–1331.

24. Fred Fishel, Jason Ferrell, Greg MacDonald, and Brent Sellers, "Herbicides: How Toxic Are They?," Document PI-133, UF/IFAS Extension, Institute of Food and Agricultural Sciences, University of Florida, September 2006, rev. February 2013, https://edis.ifas.ufl.edu/pi170.

25. Sales sheet for Roundup Ready, Monsanto Company, St. Louis, MO, http://www.cdms.net/LDat/ld178006.pdf.

26. Nora Benachour and Giles-Eric Séralini, "Glyphosate Formulations Induce Apotosis and Necrosis in Human Umbilical, Embryonic, and Placental Cells," *Chemical Research in Toxicology* 22, no. 1 (2009): 97–105.

27. Nicolas Defarge, Eszter Takács, Verónica Laura Lozano, Robin Mesnage, Joël Spiroux de Vendômois, Gilles-Eric Séralini, and András Székács, "Co-formulants in Glyphosate-Based Herbicides Disrupt Aromatase Activity in Human Cells below Toxic Levels," *International Journal of Environmental Research in Public Health* 13, no. 3 (2016): 264.

28. Wes Maxwell, "Monsanto's Glyphosate Wiping out Monarch Butterflies; Population down 74 Percent in California," *Natural News*, July 26, 2016, https://www.naturalnews.com/054775_monarch_butterflies_milkweed_glyphosate.html.

29. John M. Pleasants and Karen S. Oberhauser, "Milkweed Loss in Agricultural Fields Because of Herbicide Use: Effect on the Monarch Butterfly Population," *Insect Conservation and Diversity* 6 (2013): 135–144.

30. International Agency Research on Cancer (IARC), "Glyphosate," in *Some Organophosphate Insecticides and Herbicides*, vol. 112 of the IARC Monographs on the Evaluation of Carcinogenic Risks to Humans, World Health Organization, August 11, 2016, http://monographs.iarc.fr/ENG/Monographs/vol112/mono112-10.pdf.

31. Shahla Hosseini Bai and Steven M. Ogbourne, "Glyphosate: Environmental Contamination, Toxicity and Potential Risks to Human Health via Food Contamination," *Environmental Science and Pollution Research* 23, no. 19 (2016): 18,988–19,001.

32. Robin Mesnage, George Renney, Gilles-Eric Séralini, Malcolm Ward, and Michael N. Antoniou, "Multiomics Reveal Non-alcoholic Fatty Liver Disease in Rats Following Chronic Exposure to an Ultra-low Dose of Roundup Herbicide," *Nature Scientific Reports* 7 (2017): 39,328–39,343.

Chapter 6: Disease-Resistant Transgenic Crops

1. Richard Broglie and Karen Broglie, "Production of Disease Resistant Transgenic Plants," *Current Opinion in Biotechnology* 2 (1993): 148.

2. Uwe Conrath, "Systemic Acquired Resistance," *Plant Signaling and Behavior* 1, no. 4 (2006): 179–184.

3. J. Fletcher, C. Bender, B. Budowle, W. T. Cobb, S. E. Gold, et al., "Plant Pathogen Forensics: Capabilities, Needs, and Recommendations," *Microbiological and Molecular Biology Reviews* 70, no. 2 (2006): 450–471.

4. Ronald J. Howard, J. Allan Garland, and W. Lloyd Seaman, "Crop Losses and Their Causes," in *Diseases and Pests of Vegetable Crops in Canada* (Ottawa, CA: The Canadian Phytopathological Society and the Entomological Society of Canada, 1994), 67–78, https://phytopath.ca/wp-content/uploads/2015/03/DPVCC-Chapter-2-Crop -losses.pdf.

5. Society for General Microbiology, "Combatting Plant Diseases Is Key for Sustainable Crops," *Science Daily*, April 13, 2011, https://www.sciencedaily.com/releases /2011/04/110411194819.htm.

6. D. D. Jensen, "Papaya Virus Diseases with Special Reference to Papaya Ringspot," *Phytopathology* 39 (1949): 191–211.

7. Jon Y. Suzuki, S. Tripathi, and D. Gonsalves, "Virus-Resistant Transgenic Papaya: Commercial Development and Regulation and Environmental Issues," in *Biotechnology and Plant Disease Management*, ed. Zamir K. Punja, Solke H. De Boer, and Helene I. Sanfaçon, chap. 19 (Wallingford, UK: Centre for Agriculture and Biosciences, 2007).

8. Ronald A. Heu, Norman M. Nagata, Mach T. Fukada, and Bob Y. Yonahara, "Papaya Ringspot Virus Established on Maui," New Pest Advisory No. 02-03, Hawaii Department of Agriculture, May 2002, https://hdoa.hawaii.gov/pi/files/2013/01/npa 02-03_prvmaui.pdf.

9. H. L. Wang, S. D. Yeh, R. J. Chiu, and D. Gonsalves, "Effectiveness of Cross Protection by Mild Mutants of Papaya Ringspot Virus for Control of Ringspot Disease of Papaya in Taiwan," *Plant Disease* 71 (1987): 491–497.

10. S. D. Yeh and D. Gonsalves, "Practices and Perspectives of Control of Papaya Ringspot Virus by Cross Protection," *Advances in Disease Vector Research* 10 (1994): 237–257.

11. S. D. Yeh, J. A. J. Raja, Y. J. Kung, and W. Kositratana, "Agbiotechnology: Costs and Benefits of Genetically Modified Papaya," in *Encyclopedia of Agriculture and Food Systems*, ed. Neal K. Van Alfen, 35–50 (London: Elsevier, 2014).

12. P. P. Abel, R. S. Nelson, B. De, N. Hoffman, S. G. Rogers, R. T. Fraley, and R. N. Beachy, "Delay of Disease Development in Transgenic Plants That Express the Tobacco Mosaic Virus Coat Protein Gene," *Science* 232 (1986): 736–743.

13. Yeh et al., "Agbiotechnology," 47.

14. Marc Fuchs and Dennis Gonsalves, "Safety of Virus-Resistant Transgenic Plants Two Decades after Their Introduction: Lessons from the Realistic Field Assessment Studies," *Animal Review of Phytopathology* 45 (2007): 183.

15. Fuchs and Gonsalves, "Safety of Virus-Resistant Transgenic Plants," 183.

16. C. Neal Stewart Jr., Matthew D. Halfhill, and Suzanne I Warwick, "Transgence Introgression from Genetically Modified Crops to Their Wild Relatives," *Nature* 4 (2003): 806.

17. Fuchs and Gonsalves, "Safety of Virus-Resistant Transgenic Plants," 187.

18. Zhe Jiao, Jianchao Deng, Gongke Li, Zhuomin Zhang, and Zongwei Cai, "Study on the Compositional Differences between Transgenic and Non-transgenic Papaya (*Carica papaya* L.)," *Journal of Food Composition and Analysis* 23 (2010): 640–647.

19. Xiang-Dong Wei, Hui-Ling Zou, Lee-Min Chu, Bin Liao, Chang-Min Ye, and Chong-Yu Lan, "Field Released Transgenic Papaya Affects Microbial Communities and Enzyme Activities in Soil," *Plant Soil* 285 (2006): 347–358.

20. Wei et al., "Field Release Transgenic Papaya," 335.

21. Wei et al., 356.

22. Wei et al., 348.

23. Md. Abul Kalam Azad, Latifah Amin, and Nik Marzuki Sidik, "Gene Technology for Papaya Ringspot Virus Disease Management," *Scientific World Journal* 2014 (2014): 1–11.

24. W. Shen, G. Yang, Y. Chen, P. Yan, D. Tuo, X. Li, and P. Zhou, "Resistance of Non-transgenic Papaya Plants to Papaya Ringspot Virus (PRSV) Mediated by Intron-Containing Hairpin dsRNAs Expressed in Bacteria," *ACTA Virologica* 58 (2014): 262.

Chapter 7: Insect-Resistant Crops

1. Allah Bakhsh, Saber D. Khabbazi, Faheem S. Baloch, et al., "Insect-Resistant Transgenic Crops: Retrospect and Challenges," *Turkish Journal of Agriculture and Forestry* 39 (2015): 1–18.

2. Robert A. Zakharyan, A. Israelyaynu, A. S. Agabalyan, E. Tatevosyapn, M. Akopyans, and K. Afrikyane, "Plasmid DNA from *Bacillus thuringiensis*," *Microbiology* 48(2) (1979): 226–229.

3. S. S. Gill, E. A, Cowles, and P. V. Pietrantomo, "The Mode of Action of *Bacillus thuringiensis* Endotoxins," *Annual Review of Entomology* 37 (1992): 615–636.

4. Davi F. Farias, Ad A. C. M. Peijnenburg, Maria S. Grossi-de-Sa, and Ana F.U. Carvalho, "Food Safety Knowledge on the Bt Mutant Protein Cry8ka5 Employed in the Development of Coleopteran-Resistant Transgenic Cotton Plants," *Bioengineered* 6, no. 6 (2015): 323–327.

5. Anthony M. Shelton, Jian-Zhou Zhao, and Richard T. Roush, "Economic, Ecological, Food Safety, and Social Consequences of the Development of Bt Transgenic Plants," *Annual Review of Entomology* 47 (2002): 845–881.

6. Shelton, Zhao, and Roush, "Economic, Ecological, Food Safety, and Social Consequences of the Deployment of Bt Transgenic Plants," 845.

7. Kun Xue, Jing Yang, Biao Liu, and Day van Xue, "The Integrated Risk Assessment of Transgenic Rice *Oryza sativa*: A Comparative Proteomics Approach," *Food Chemistry* 135 (2012): 314.

8. Mark Vaeck, Arlette Reynaerts, Herman Höfte, et al., "Transgenic Plants Protected from Insect Attack," *Nature* 327 (1987): 33–37.

9. Luca Bucchini and Lynn R. Goldman, "StarLink Corn: A Risk Analysis," *Environmental Health Perspectives* 110, no. 11 (2002): 5–13.

10. Bucchini and Goldman, "StarLink Corn," 11.

11. Michael S. Koch, Jason W. Ward, Steven L. Levine, James A. Baum, John L. Vicini, and Bruce G. Hammond, "The Food and Environmental Safety of Bt Crops," *Frontiers in Plant Science* 6 (2015): 16.

12. Deland R. Juberg, Rod A. Herman, Johnson Thomas, Keith J. Brooks, and Bryan Delaney, "Acute and Repeated Dose (28 Days) Mouse Oral-Toxicology Studies with Cry34Ab1 and Cry35Ab1 Bt Proteins Used in Coleopteran Resistant DAS-59122-7 Corn," *Regulatory Toxicology and Pharmacology* 54 (2009): 154.

13. Nestor Rubio-Infante and Leticia Moreno-Fierros, "An Overview of the Safety and Biological Effects of *Bacillus thuringiensis* Cry Toxins in Mammals," *Journal of Applied Toxicology* 36 (2016): 630–648.

14. Aysun Kilic and M. Turan Akay, "A Three Generation Study with Genetically Modified Bt Corn in Rats: Biochemical and Histopathological Investigation," *Food and Chemical Toxicology* 46 (2008): 1164–1170.

15. Michael N. Antoniou and Claire J. Robinson, "Cornell Alliance for Science Evaluation of Consensus on Genetically Modified Food Safety: Weaknesses in Study Design," *Frontiers in Public Health* 5 (April 13, 2017).

16. Xue et al., "The Integrated Risk Assessment of Transgenic rice *Oryza sativa*," 317.

17. Z. R. Akhtar, J. C. Tian, Y. Chen, Q. Fang, C. Hu, Y. F. Peng, and G. Y. Ye, "Impact of Six Transgenic *Bacillus thuringiensis* Rice Lines on Four Nontarget Thrips Species Attacking Rice Panicles in the Paddy Field," *Transgenic Plants and Insects* 42, no. 1 (2013): 173–180.

18. Miguel A. Altieri, "The Myth of Coexistence: Why Transgenic Crops Are Not Compatible with Agroecologically Based Systems of Production," *Bulletin of Science, Technology and Society* 25, no. 4 (2005): 361–371.

19. Farias et al., "Food Safety Knowledge on the Bt Mutant Protein," 326.

Chapter 8: Genetic Mechanisms and GMO Risk Assessment

1. Li Patrushev and T. F. Kovalenko, "Functions of Noncoding Sequences in Mammalian Genomes," *Biochemistry* (Moscow) 79, no. 13 (2014): 1442–1469.

2. Evelyn Fox Keller, *The Century of the Gene* (Cambridge, MA: Harvard University Press, 2000), 146.

3. Evan Charney, "Review Essay: Sheldon Krimsky and Jeremy Gruber, eds., *Genetic Explanations: Sense and Nonsense*," *Logos* 12 , no. 3 (2013).

4. Loren Graham, *Lysenko's Ghost: Epigenetics and Russia* (Cambridge, MA: Harvard University Press, 2016).

5. Leslie Pray and Kira Zhaurova, "Barbara McClintock and the Discovery of Jumping Genes (Transposons)," *Nature Education* 1, no. 1 (2008): 169, http://www.nature.com/scitable/topicpage/barbara-mcclintock-and-the-discovery-of-jumping-34083.

6. Alternative splicing starts with the same DNA sequence. One process of protein synthesis might include exons that are excluded in another, which produces different messenger RNA molecules that result in a different configuration of amino acids and thus a different protein molecule. Alternative splicing gives plasticity to the human genome, allowing about 20,000 coding genes to produce about 100,000 proteins.

7. A homologous transgene is an artificial cassette that is flanked by regions that are homologous to sequences bordering the targeted gene. It is used as a research technique to inactivate a gene and determine its function in a living animal.

8. H. Vaucheret, C. Béclin, T. Elmayanet, F. Feuerbach, C. Godon, J. B. Morel, P. Mourrain, J. C. Palauqui, and S. Vernhettes, "Transgene-Induced Gene Silencing in Plants," *Plant Journal* 16, no. 6 (1998): 651, 656.

9. David Schubert, "A Different Perspective on GM Food," *Nature Biotechnology* 20 (October 2002): 969.

10. Kent J. Bradford, Allen Van Deynze, Neal Gutterson, Wayne Parrott, and Steven H. Strauss, "Regulating Transgenic Crops Sensibly: Lessons from Plant Breeding, Biotechnology and Genomics," *Nature Biotechnology* 23 (2005): 439.

11. Nina Federoff and Nancy Marie Brown, *Mendel in the Kitchen: A Scientist's View of Genetically Modified Foods* (Washington, DC: Joseph Henry Press, 2004), 165.

12. Institute of Medicine and National Research Council, *Safety of Genetically Engineered Foods: Approaches to Assessing Unintended Health Effects* (Washington, DC: National Academies Press, 2004), 8.

13. George Acquaah, *Principles of Plant Genetics and Breeding* (Oxford, UK: Wiley-Blackwell, 2012), 445.

14. National Academies of Sciences, Engineering, and Medicine, *Genetically Engineered Crops: Experiences and Prospects* (Washington, DC: National Academies Press, 2016), 66.

15. R. Beachy, J. L. Bennetzen, B. M. Chassy et al., "Letter to the Editor on 'Divergent Perspectives on GM Food,'" *Nature Biotechnology* 20, no. 12 (December 2002): 1195–1196.

16. Jonathan R. Latham, Allison K. Wilson, and Ricarda A. Steinbrecher, "The Mutational Consequences of Plant Transformation," *Journal of Biomedicine and Biotechnology* 206 (November 22, 2005): 1–7.

17. Brian Wynne, "Reflexing Complexity," *Theory, Culture and Society* 22, no. 5 (2005): 67–94.

18. Hae-Woon Choi, Peggy Lemaux, and Myeong-Je Cho, "High-Frequency Cytogenic Aberration in Transgenic Oat," *Plant Science* 160 (2001): 763–772.

19. Federoff and Brown, *Mendel in the Kitchen*, 174.

20. Federoff and Brown, 174.

21. Claudia Paoletti, Eric Flamm, William Yan, Sue Meek, Suzy Renkens, Marc Fellous, and Henry Kuiper, "GMO Risk Assessment around the World: Some Examples," *Trends in Food Science and Technology* 19 (2008): s71.

22. Andrew Cockburn, "Assuring the Safety of Genetically Modified (GM) Foods: The Importance of a Holistic Integrative Approach," *Journal of Biotechnology* 98 (2002): 93.

23. Organisation for Economic Co-operation and Development, *Food Safety Evaluation*, Report of Workshop on Food Safety Evaluation held in Oxford, UK, September 12–15, 1994 (Paris: OECD, 1996).

24. Andrew Bartholomaeus, Wayne Parrott, Genevieve Bondy, and Kate Walker, "The Use of Whole Food Animal Studies in the Safety Assessment of Genetically Modified Crops: Limitations and Recommendations," *Critical Reviews in Toxicology* 43, no. 52 (2013): 19.

25. Bartholomaeus, Parrott, Bondy, and Walker, "The Use of Whole Food Animal Studies," 19.

26. Alexander Y. Panchin and Alexander I. Tuzhikov, "Published GMO Studies Find No Evidence of Harm When Corrected for Multiple Comparisons," *Critical Reviews in Biotechnology* 37, no. 2 (2017): 216.

27. Heiko Rischer and Kirsi-Marja Oksman-Caldentey, "Unintended Effects in Genetically Modified Crops: Revealed by Metabolmics?," *Trends in Biotechnology* 24, no. 3 (2006): 102.

Chapter 9: Contested Viewpoints on the Health Effects of GMOs

1. Andrew Cockburn, "Assuring the Safety of Genetically Modified (GM) Foods: The Importance of a Holistic, Integrative Approach," *Journal of Biotechnology* 98 (2002): 83.

2. Andrew Pollack, "U.S.D.A. Approves Modified Potato. Next Up: French Fry Fans," *New York Times*, November 7, 2014.

3. John Fagan, Michael Antoniou, and Claire Robinson, *GMO Myths and Truths*, 2nd ed. (London: Earth Open Source, 2014), http://earthopensource.org/earth-open -source-reports/gmo-myths-and-truths-2nd-edition.

4. Council of the European Communities, Council Directive of April 23, 1990 on the Deliberate Release into the Environment of Genetically Modified Organisms, 90/220/EEC.

5. Cockburn, "Assuring the Safety of Genetically Modified (GM) Foods," 85.

6. Jonathan R. Latham, Allison K. Wilson, and Ricarda A. Steinbrecher, "The Muta-tional Consequences of Plant Transformation," *Journal of Biomedicine and Biotechnol-ogy* (2006): 1–7.

7. Nina Federoff and Nancy Marie Brown, *Mendel in the Kitchen: A Scientist's View of Genetically Modified Foods* (Washington, DC: Joseph Henry Press, 2004), 174.

8. Latham, Wilson, and Steinbrecher, "The Mutational Consequences of Plant Trans-formation," 4.

9. Harry A. Kuiper, Gijs A. Kleter, P. Hub, J. M. Noteborn, and Esther J. Ko, "Assess-ment of the Food Safety Issues Related to Genetically Modified Foods," *Plant Journal* 27, no. 6 (2001): 503–528.

10. Federoff and Brown, *Mendel in the Kitchen*, 175.

11. Cockburn, "Assuring the Safety of Genetically Modified (GM) Foods," 93.

12. Organisation for Economic Co-operation and Development, *Agricultural Policies in OECD Countries: Monitoring and Evaluation 2000* (Cedex, France: OECD, June 2000), https://www.oecd-ilibrary.org/docserver/agr_oecd-2000-en.pdf?expires=1525810555 &id=id&accname=oid006278&checksum=5BA0545A667750F8E2A945394B8E7A7B.

13. Erik Millstone, Eric Brunner, and Sue Mayer, "Beyond 'Substantial Equivalence,'" *Nature* 401 (1999): 525–526.

14. Cockburn, "Assuring the Safety of Genetically Modified (GM) Foods," 94.

15. Fagan, Antoniou, and Robins, *GMO Myths and Truths*, sec. 2.1, 62.

16. Paul Lurquin, *High Tech Harvest: Understanding Genetically Modified Food Plants* (Boulder, CO: Westview Press, 2002), 99.

17. Athenix Corporation, "Early Food Safety Evaluation for EPSPS ACES Protein," submission to the Office of Food Additive Safety, Division of Biotechnology and GRAS Notice Review, HFS-255, Center for Food Safety and Applied Nutrition, Food and Drug Administration, October 7, 2009, http://www.fda.gov/downloads/Food/Biotech nology/Submissions/UCM233624.pdf.

18. Lurquin, *High Tech Harvest*, 102.

19. Cockburn, "Assuring the Safety of Genetically Modified (GM) Foods," 89.

20. Qixing Mao, Fabiana Manservisi, Simona Panzacchi, Daniele Mandrioli, Ilaria Menghetti, Andrea Vornoli, Luciana Bua, et al. "The Ramazzini Institute 13-Week Pilot Study on Glyphosate Administered at Human-Equivalent Dose to Sprague-Dawley Rats: Effects on the Microbiome," *Environmental Health* 17, no. 1 (May 2018), https://glyphosatestudy.org/wp-content/uploads/2018/05/MICROBIOME-GLY-PILOT -IN-PRESS-8-5-1.pdf.

21. Suzie Key, Julian K.-C. Ma, and Pascal M. W. Drake, "Genetically Modified Plants and Human Health," *Journal of the Royal Society of Medicine* 101, no. 6 (2008): 292.

22. Carey Gillem, "New Claims against Monsanto in Consumer Lawsuit over Roundup Herbicide," *Huffington Post*, June 20, 2017.

23. Alessandro Nicolia, Alberto Manzo, Fabio Veronesi, and Daniele Rosellini, "An Overview of the Last Ten Years of Genetically Engineered Crop Safety Research," *Critical Review of Biotechnology* 34, no. 1 (2014): 77–88.

24. Alexander Y. Panchin and Alexander I. Tuzhikov, "Published GMO Studies Find No Evidence of Harm When Corrected for Multiple Comparisons," *Critical Reviews in Biotechnology* 37, no. 2 (2017): 213–217.

25. Andrew Bartholomaeus, Wayne Parrott, Genevieve Bondy, and Kate Walker, "The Use of Whole Food Animal Studies in the Safety Assessment of Genetically Modified Crops: Limitations and Recommendations," *Critical Reviews in Toxicology* 43, no. 52 (2013): 1–24.

26. Federoff and Brown, *Mendel in the Kitchen*.

27. Katy L. Johnson, Alan F. Raybould, Malcolm D. Hudson, and Guy M. Poppy, "How Does Scientific Risk Assessment of GM Crops Fit within the Wider Risk Analysis," *Trends in Plant Science* 12, no. 1 (2006): 1–5.

28. U.S. Food and Drug Administration, "Statement of Policy: Foods Derived from New Plant Varieties," *Federal Register* 57, no. 104 (1992): 22,991.

29. Fagan, Antoniou, and Robinson, *GMO Myths and Truths*, sec. 2.2.

30. Tomoko Inose and Kousaku Murata, "Enhanced Accumulation of Toxic Compound in Yeast Cells Having High Glycolytic Activity: A Case Study on the Safety of Genetically Engineered Yeast," *International Journal of Food Science and Technology* 30 (1995): 141–146.

31. Fagan, Antoniou, and Robinson, *GMO Myths and Truths*, sec. 2.2.

32. David A. Kessler, Michael R. Taylor, James H. Maryanski, Eric L. Flamm, and Linda S. Kahl, "The Safety of Foods Developed by Biotechnology," *Science* 256 (1992): 1832.

33. Fagan, Antoniou, and Robinson, *GMO Myths and Truths*, sec. 2.2.

34. Johnson, Raybould, Hudson, and Poppy, "How Does Scientific Risk Assessment of GM Crops Fit within the Wider Risk Analysis," *Trends in Plant Science* 12, no. 1 (2006): 1–5.

Chapter 10: Labeling GMOs

1. Jason Kelly, "I Run a G.M.O. Company—and I Support G.M.O Labeling," *New York Times*, op-ed, May 16, 2016.

2. U.S. Food and Drug Administration, "Guidance to Industry for Foods Derived from New Plant Varieties," *Federal Register* 57, no. 104 (May 29, 1992): 22,291.

3. U.S. Food and Drug Administration, "Section 601.53. Submission of Certain Data and Information Related to Human Gene Therapy or Xenotransplantation for Public Disclosure," *Federal Register* 66, no. 12 (January 18, 2001): 4,706–4,738.

4. Alliance for Bio-Integrity v. Shalala, 116 F. Supp. 2d 166 (D.C.C. 2000), http://law .justia.com/cases/federal/district-courts/FSupp2/116/166/2576171.

5. An Act Concerning Genetically-Engineered Food, Connecticut Public Act No. 13-183, October 1, 2013.

6. An Act to Protect Maine Food Consumers' Right to Know about Genetically Engineered Food and Seed Stock, Maine Law LD 718, February 26, 2013.

7. An Act Relating to the Labeling of Food Produced with Genetic Engineering, Vermont Act No. 120, May 8, 2014.

8. Dave Gram, "Vermont Law on GMO Food to Stand for Now," *Burlington Press*, April 28, 2015.

9. U.S. Government Printing Office, H.R. 1599: Safe and Accurate Food Labeling Act of 2015, July 24, 2015, https://www.gpo.gov/fdsys/pkg/BILLS-114hr1599rfs/pdf /BILLS-114hr1599rfs.pdf.

10. National Bioengineered Food Disclosure Standard, S.764, June 23, 2016, https://www.congress.gov/bill/114th-congress/senate-bill/764.

11. Courtney Begley, "So Close Yet So Far: The United States Follows the Lead of the European Union in Mandating GMO Labeling. But Did It Go Far Enough?," *Fordham International Law Journal* 40 (February 2017): 625–727.

12. Jonathan H. Adler, "Compelled Commercial Speech and the Consumer 'Right to Know,'" *Arizona Law Review* 58 (2016): 421.

13. Charlotte Davis, "A Right to Know about GMOs: What *American Meat Institute v. USDA* Means for Vermont's Food Labeling Law," *North Carolina Journal of Law and Technology* 16 (April 2015): 32.

14. Cheryl Hogue, "Industry Takes Aim at State Chemical Disclosure Laws, *Chemical & Engineering News* 96, no. 12 (March 2018): 21–22.

15. Julie M. Muller, "Naturally Misleading: FDA's Unwillingness to Define 'Natural' and the Quest for GMO Transparency through State Mandatory Labeling Initiatives," *Suffolk University Law Journal* 48 (2015): 519.

16. Mark Bittman, "G.M.O. Labeling Could Stir Up a Revolution," *New York Times*, op-ed, September 2, 2016, A19.

17. Begley, "So Close Yet So Far," 664.

18. Begley, 667.

19. Jennifer McGee, "Weird Science: Frankenstein Foods and States as Laboratories," *Journal of Law and Health* 30 (2017): 147.

Chapter 11: The 2016 National Academies Study

1. National Academies of Sciences, Engineering, and Medicine (NASEM), "Who We Are," http://www.nationalacademies.org/about/whoweare/index.html.

2. National Research Council, *Agricultural Biotechnology: Strategies for National Competitiveness* (Washington, DC: National Academy Press, 1987).

3. National Research Council, *Field Testing Genetically Modified Organisms: Framework for Decisions* (Washington, DC: National Academy Press, 1989).

4. National Research Council, *Genetically Modified Pest-Protected Plants: Science and Regulation* (Washington, DC: National Academy Press, 2000).

5. National Research Council, *The Environmental Effect of Transgenic Plants: The Scope and Adequacy of Regulation* (Washington, DC: National Academies Press, 2002).

6. National Research Council, *Safety of Genetically Engineered Foods: Approaches to Assessing Unintended Health Effects* (Washington, DC: National Academies Press,

2004); National Research Council, *Biological Confinement of Genetically Engineered Organisms* (Washington, DC: National Academies Press, 2004).

7. National Research Council, *The Impact of Genetically Engineered Crops on Farm Sustainability in the United States* (Washington, DC: National Academies Press, 2010).

8. National Academies of Science, Engineering, and Medicine (NASEM), *Genetically Engineered Crops: Experiences and Prospects* (Washington, DC: National Academies Press, 2016), 36.

9. NASEM, *Genetically Engineered Crops*.

10. NASEM, 58.

11. National Research Council, *The Impact of Genetically Engineered Crops*, 100.

12. National Research Council, 135.

13. Heidi Ledford, "Brazil Considers Transgenic Trees," *Nature*, August 27, 2014, https://www.nature.com/news/brazil-considers-transgenic-trees-1.15769.

14. NASEM, *Genetically Engineered Crops*, 102.

15. NASEM, 116.

16. United Nations Department of Economic and Social Affairs, "World Population Projected to Reach 9.7 Billion by 2050," July 29, 2015, http://www.un.org/en/develop ment/desa/news/population/2015-report.html.

17. NASEM, *Genetically Engineered Crops*, 154.

18. Khuda Bak, "Impacts of Bt Cotton on Profitability, Productivity and Farm Inputs in Pakistan: Use of Panel Models," *Environment and Development Economics* 22, no. 4 (2017): 373–391, doi.org/10.1017/s1355770x17000080.

19. NASEM, *Genetically Engineered Crops*, 173.

20. NASEM, 175.

21. NASEM, 175.

22. NASEM, 176.

23. NASEM, 178.

24. NASEM, 185.

25. National Research Council, *The Environmental Effect of Transgenic Plants*, 15.

26. NASEM, *Genetically Engineered Crops*, 191.

27. Gilles-Eric Séralini, Robin Mesnage, Nicolas Defarge, Steeve Gress, Didier Hennequin, Emilie Clair, Manuela Malatesta, and Joël Spiroux de Vendômois, "Answers

to Critics: Why There Is a Long Term Toxicity Due to a Roundup-Tolerant Genetically Modified Maize and to a Roundup Herbicide," *Food and Chemical Toxicology* 53 (2013): 476–483.

28. European Food Safety Authority, "Review of the Séralini et al. (2012) Publication on a 2-Year Rodent Feeding Study with Glyphosate Formulations and GM Maize NK603 as Published Online on 19 September 2012 in *Food and Chemical Toxicology*," *EFSA Journal* 10, no. 10 (2012): 2910.

29. Gilles-Eric Séralini, Robin Mesnage, Nicolas Defarge, and Joël Spiroux de Vendômois, "Conclusiveness of Toxicity Data and Double Standards," *Food and Chemical Toxicology* 69 (2014): 357.

30. NASEM, *Genetically Engineered Crops*, 197.

31. European Food Safety Authority, "Statistical Significance and Biological Relevance," *EFSA Journal* 9, no. 9 (2011): 2372–2389.

32. Michael F. W. Festing and Douglas G. Altman, "Guidelines for the Design and Statistical Analysis of Experiments Using Laboratory Animals," *ILAR Journal* 43, no. 4 (2002): 244–258.

33. Séralini et al., "Answers to Critics," 482.

34. NASEM, *Genetically Engineered Crops*, 194.

35. European Food Safety Authority, "Statistical Significance and Biological Relevance," 2385.

36. Aysun Kiliç and Mehmet Turan Akay, "A Three-Generation Study with Genetically Modified Bt Corn in Rats: Biochemical and Histopathological Investigation," *Food and Chemical Toxicology* 46, no. 3 (2008): 1164–1170.

37. Kiliç and Akay, "A Three-Generation Study," 1169.

38. NASEM, 194.

39. NASEM, 195.

40. NASEM, 195.

41. NASEM, 195.

42. Alison L. Van Eenennaam and Andrea E. Young, "Prevalence and Impacts of Genetically Engineered Feedstuffs on Livestock Populations," *Journal of Animal Science* 92, no. 10 (2014): 4255–4278.

43. NASEM, *Genetically Engineered Crops*, 195.

44. NASEM, 198.

45. NASEM, 226.

46. NASEM, 236.

47. Sheldon Krimsky, "An Illusory Consensus behind GMO Health Assessment," *Science, Technology and Human Values* 40, no. 6 (2015): 883–914.

48. A. Dona and I. S. Arvanitouannis, "Health Risks of Genetically Modified Foods," *Critical Reviews in Food Science and Nutrition* 49, no. 2 (2009): 164–175.

49. Chelsea Snell, Aude Bernheim, Jean-Baptiste Berge, Marcel Kuntz, Gérard Pascal, Alain Paris, and Agnès E. Ricroch, "Assessment of the Health Impact of GM Plant Diets in Long-Term and Multigenerational Animal Feeding Trials: A Literature Review," *Food and Chemical Toxicology* 50, no. 3–4 (2012): 1134–1148.

50. Federal Advisory Committee Act of 1997, sec.15(b).

51. Merrill Goozner and Corrie Maudlin, "Ensuring Independence and Objectivity at the National Academies," Center for Science in the Public Interest, Washington, DC, July 1, 2006, https://cspinet.org/new/pdf/nasreport.pdf.

52. Sheldon Krimsky and Tim Schwab, "Conflicts of Interest among Committee Members in the National Academies' Genetically Engineered Crop Study," *PLOS ONE*, February 26, 2017, doi: 10.1371/journal.pone.0172317.

53. National Academies of Sciences, Engineering, and Medicine (NASEM), "Statement by the National Academies of Sciences, Engineering, and Medicine regarding *PLOS ONE* Article on Our Study of Genetically Engineered Crops," news release, March 1, 2017.

54. Ashley P. Taylor, "National Academies Revise Conflict of Interest Policy," *The Scientist*, May 3, 2017, http://www.the-scientist.com/?articles.view/articleNo/49331/title/National-Academies-Revise-Conflict-of-Interest-Policy.

Chapter 12: The Promise and Protests of Golden Rice

1. Peter Beyer, "Golden Rice and 'Golden' Crops for Human Nutrition," *New Biotechnology* 27, no. 5 (2010): 478–481.

2. Martin Ensernik, "Tough Lessons from Golden Rice," *Science* 320 (2008): 468–471.

3. Ingo Potrykus, "Golden Rice and Beyond," *Plant Physiology* 125 (2001): 1158.

4. Marion Nestle, *Safe Food: The Politics of Food Safety* (Berkeley: University of California Press, 2010), 153.

5. Nestle, *Safe Food*, chap. 5.

6. Xudong Ye, Salim Al-Babili, Andreas Klöti, Jing Zhang, Paola Lucca, Peter Beyer, and Ingo Potrykus, "Engineering the Provitamin A (β-carotene) Biosynthetic Pathway into (Caretenoid-Free) Rice Endosperm," *Science* 287, no. 5451 (2000): 303–305;

Jacqueline A. Paine et al., "Improving the Nutritional Value of Golden Rice through Increased Pro-vitamin A Content," *Nature Biotechnology* 23, no. 4 (2005): 482–487.

7. Mae-Wan Ho and Joe Cummins, "The 'Golden Rice': An Exercise in How Not to Do Science," Institute of Science in Society, London, June 13, 2000, http://www.i-sis .org.uk/rice.php.

8. X. Ye, S. Al-Babili, J. Zhang, P. Lucca, P. Beyer, and I. Potrykus, "Engineering the Provitamin A (Beta-carotene) Biosynthetic Pathway into (Carotenoid-Free) Rice Endosperm," *Science* 5451 (2000): 303–305.

9. Peter Beyer, Salim Al-Babili, Xudong Ye, Paola Lucca, Patrick Schaub, Ralf Welsch, and Ingo Potrykus, "Golden Rice: Introducing the β-carotene Biosynthesis Pathway into Rice Endosperm by Genetic Engineering to Defeat Vitamin A Deficiency," *Journal of Nutrition* 132, no. 3 (2002): 506S–510S.

10. Guangwen Tang, Jian Qin, Gregory G. Dolnnikowski, Robert M. Russell, and Michael A. Grusak, "Golden Rice Is an Effective Source of Vitamin A," *American Journal of Clinical Nutrition* 89 (2009): 1776–1783.

11. Guangwen Tang, Yuming Hu, Shi-an Yin, Yin Wang, Gerard E. Dallal, Michael A. Grusak, and Robert M. Russell, "β-carotene in Golden Rice Is as Good as β-carotene in Oil at Providing Vitamin A to Children," *American Journal of Clinical Nutrition* 96, no. 3 (2012): 658–664.

12. Xiang Yu and Wei Li, "Informed Consent and Ethical Review in Chinese Human Experimentation: Reflections on the 'Golden Rice Event,'" *Biotechnology Law Report* 33 (2014): 155–160.

13. Yu and Li, "Informed Consent."

14. Guangwen Tang, Jian Qin, Gregory G. Dolnikowski, Robert M Russell, and Michael A. Grusak, "Golden Rice Is an Effective Source of Vitamin A," *American Journal of Clinical Nutrition* 89 (2009): 1176–1183.

15. "Patents for Humanity Awards 2015," U.S. Patent and Trademark Office, 2015, https://www.uspto.gov/patent/initiatives/patents-humanity/patents-humanity-awards -2015.

16. Ingo Potrykus, "The 'Golden Rice' Tale," *In Vitro Cell Developmental Biology Plant* 37 (2011): 93–100.

17. Joel Achenbach, "107 Nobel Laureates Sign Letter Blasting Greenpeace over GMOs," *Washington Post*, June 30, 2016.

18. Mae-Wan Ho and Joe Cummins, "The 'Golden Rice: Scandal Unfolds'" *Science & Society* 42 (Summer 2009), http://www.i-sis.org.uk/isisnews/sis42.php.

19. Ensernik, "Tough Lessons from Golden Rice," 468.

20. Justus Wesseler and David Zilberman, "The Economic Power of the Golden Rice Opposition," *Environment and Development Economics* 19 (2013): 724–742.

21. Hyejin Lee and Sheldon Krimsky, "The Arrested Development of Golden Rice: The Scientific and Social Challenges of a Transgenic Biofortified Crop," *International Journal of Social Science Studies* 4, no. 11 (2016): 51–64.

22. International Rice Research Institute (IRRI), home page, http://irri.org.

23. American Society for Nutrition, "Researchers Determine That Golden Rice Is an Effective Source of Vitamin A," June 8, 2009, http://www.goldenrice.org/PDFs/ASN onGR.pdf.

24. Guangwen Tang, "Bioconversion of Dietary Provitamin A Carotenoids to Vitamin A in Humans," *American Journal of Clinical Nutrition* 91 (supp.) (2010): 1468S–143S.

25. International Service for the Acquisition of Agri-biotech Applications, "Biotechnology and Biofortification," Pocket K No. 27, ISAAA, June 2007, http://www.isaaa .org/resources/publications/pocketk/27/default.asp.

26. Kathleen L. Hefferon, "Nutritionally Enhanced Food Crops: Progress and Perspectives," *International Journal of Molecular Sciences* 16 (2015): 3895–3914.

27. Hans De Steur, Joshua Wesana, Dieter Blancquaert, Dominique Van Der Straeten, and Vavier Gellynick, "The Socioeconomics of Genetically Modified Biofortified Crops: A Systematic Review and Meta-analysis," *Annals of the New York Academy of Sciences* 1390 (2017): 14–33.

28. Hyejin Lee, "Transgenic Pro-vitamin A Biofortified Crops for Improving Vitamin A Deficiency and Their Challenges," *Open Agricultural Journal* 11 (2017): 3–15.

Chapter 13: Science Studies and the GMO Conflict

1. Steven Yearley, "Nature and the Environment: Science and Technology Studies," in *The Handbook of Science and Technology Studies*, 3rd ed., ed. Edward J. Hackett, Olga Amsterdamska, Michael E. Lynch, and Judy Wajcman (Cambridge, MA: MIT Press, 2008), 921–947 at 937.

2. Amaranta Herrero, Fern Wickson, and Rosa Binimelis, "Seeing GMOs from a Systems Perspective: The Need for Comparative Cartographies of Agri/Cultures for Sustainability Assessment," *Sustainability* 7 (2015): 11322.

3. Sheila Jasanoff, *Designs on Nature: Science and Democracy in Europe and the United States* (Princeton, NJ: Princeton University Press, 2005), 290.

4. Steven Yearley, "Mapping and Interpreting Societal Responses to Genetically Modified Crops and Food," *Social Studies of Science* 31, no. 1 (2001): 151–160.

5. Marjolein B. A. van Asselt and Ellen Vos, "Wrestling with Uncertain Risks: EU Regulation of GMOs and the Uncertainty Paradox," *Journal of Risk Research* 11, no. 1–2 (2008): 281–300.

6. Sheldon Krimsky and Dominic Golding, *Social Theories of Risk* (Westport, CT: Praeger, 1992).

7. Renata Motta, "Social Disputes over GMOs: An Overview," *Sociology Compass* 8, no. 12 (2014): 1360–1376.

8. Ronald Herring and Robert Paarlberg, "The Political Economy of Biotechnology," *Annual Review of Resource Economics* 8 (2016): 410.

9. Herring and Paarlberg, "The Political Economy of Biotechnology," 411.

10. Philip McMichael, "A Food Regime Genealogy," *Journal of Peasant Studies* 36, no. 1 (2009): 142.

11. McMichael, "A Food Regime Genealogy," 141.

12. Monsanto, "Monsanto Technology/Stewardship Agreement," 2011, https://the farmerslife.files.wordpress.com/2012/02/scan_doc0004.pdf.

13. Herring and Paarlberg, "The Political Economy of Biotechnology," 400.

14. Vandana Shiva, *Biopiracy: The Plunder of Nature and Knowledge* (Boston: South End Press, 1997), 33.

15. Patent Act of 1790, 1 Stat. 109-112, chap. 7, http://www.ipmall.info/sites/default /files/hosted_resources/lipa/patents/Patent_Act_of_1790.pdf.

16. R. Stephen Crespi, "An Analysis of Moral Issues Affecting Patenting Inventions in the Life Sciences: A European Perspective," *Science and Engineering Ethics* 6 (2000): 159.

17. Jerry Cayford, "Resources for the Future," letter, *Nature Biotechnology* 21 (2003): 493.

18. Jerry Cayford, "Resources for the Future."

19. Crespi, "An Analysis of Moral Issues Affecting Patenting Inventions in the Life Sciences."

20. Philippe Sands, *Principles of International Environmental Law* (Cambridge, UK: Cambridge University Press, 1995), 1048.

21. Peter Newell, "Bio-hegemony: The Political Economy of Agricultural Biotech-nology in Argentina," *Journal of Latin American Studies* 41 (2009): 53.

22. Marygold Walsh-Dilley, "Localizing Control: Mendocino County and the Ban on GMOs," *Agriculture and Human Values* 26 (2009): 96.

23. Stephen B. Brush, "Genetically Modified Organisms in Peasant Farming: Social Impact and Equity," *Indian Journal of Global Legal Studies* 9 (2001): 1–27.

24. Pablo Pellegrini, "Knowledge, Identity and Ideology in Stances on GMOs: The Case of the Movimento Sem Terra in Brazil," *Science Studies* 22, no. 1 (2009): 50.

25. Johan Diels, Marie Cunlia, Cólia Monaia, Bernardo Subugosa-Madeira, and Margarido Silva, "Association of Financial or Professional Conflict of Interest to Research Outcomes on Health Risks or Nutritional Assessment Studies of Genetically Modified Products," *Food Policy* 36 (2011): 197.

26. Renata Motta, "Social Disputes over GMOs: An Overview," *Sociology Compass* 8, no. 12 (2014): 1371.

Chapter 14: Conclusion

1. Alvin M. Weinberg, "Science and Trans-Science," *Minerva* 10, no. 2 (1974): 209–222.

2. Sheldon Krimsky, "The Weight of Scientific Evidence in Policy and Law," *American Journal of Public Health* 95 (supp. 1) (2005): S129–S136.

3. National Academies of Sciences, Engineering, and Medicine (NASEM), *Genetically Engineered Crops: Experiences and Prospects* (Washington, DC: National Academies Press, 2016), 253.

4. Yanhua Tan, Xiaoping Yi, Limin Wang, Cunzhi Peng, Yong Sun, Dan Wang, Jiaming Zhang, Anping Guo, and Xuchu Wang, "Comparative Proteomics of Leaves from Phytase-Transgenic Maize and Its Non-transgenic Isogenic Variety," *Frontiers in Plant Science* 7, no. 1211 (2016): 1–14.

5. Robin Mesnage, Sarah Z. Agapito-Tenfen, Vinicius Vilperte, George Renney, Malcolm Ward, Gilles-Eric Seraline, Rubens O. Nodari, and Michael N. Antoniou, "An Integrated Multi-omics Analysis of the NK603 Roundup-Tolerant GM Maize Reveals Metabolism Disturbances Caused by the Transformation Process," *Scientific Reports* 6, no. 37855 (2016), http://doi.org/10.1038/srep37855.

6. Natasha Gilbert, "A Hard Look at GM Crops," *Nature* 497, no. 7447 (2013): 26.

7. Wilhelm Klümper and Matin Quaim, "A Meta-analysis of the Impacts of Genetically Modified Crops," *PLOS One* 9, no. 11 (2014): 1.

8. Klümper and Quaim, "A Meta-analysis of the Impacts of Genetically Modified Crops," 4.

9. Jack A. Heinemann, Melanie Massaro, Dorien S. Coray, Sarah Zanon Agapito-Tenfen, and Jiajun Dake Wen, "Sustainability and Innovation in Staple Crop

Production in the U.S. Midwest," *International Journal of Agricultural Sustainability* 12, no. 1 (2014): 76.

10. Alan B. Bennett, Cecilia Chi-Ham, Geoffrey Barrows, Steven Sexton, and David Zilberman, "Agricultural Biotechnology: Economics, Environment, Ethics and the Future," *Annual Review of Environment and Resources* 38 (2013): 257–259.

11. Charles M. Benbrook, "Trends in Glyphosate Herbicide Use in the United States and Globally," *Environmental Science Europe* 28, no. 3 (2016): 1–15.

12. Bruce E. Tabashnik, Thierry Brévault, and Yves Carrière, "Insect Resistance to Bt Crops: Lessons from the First Billion Acres," *Nature Biotechnology* 31 (2013): 510–521.

13. Edward D. Perry, Frederico Ciliberto, David Hennessy, and GianCarlo Moschini, "Genetically Engineered Crops and Pesticide Use in U.S. Maize and Soybeans," *Science Advances* 2, no. 8 (2016): 1, doi: 10.1126/sciadv.1600850.

14. A. Wendy Russell, "GMOs and Their Contexts: A Comparison of Potential and Actual Performance of GM Crops in a Local Agricultural Setting," *Geoforum* 39, no. 1 (2008): 213.

15. Douglas H. Constance, "Sustainable Agriculture in the United States: A Critical Examination of a Contested Process," *Sustainability* 2, no. 1 (2010): 48.

16. Russell, "GMOs and Their Contexts."

17. NASEM, *Genetically Engineered Crops*, 441.

18. Kathleen L. Hefferon, "Nutritionally Enhanced Food Crops: Progress and Perspectives," *International Journal of Molecular Sciences* 16 (2015): 3895.

19. Miguel A. Altieri, "The Myth of Co-existence: Why Transgenic Crops Are Not Compatible with Agroecologically Based Systems of Production," *Bulletin of Science, Technology and Society* 25, no. 4 (2005): 369.

20. Natasha Gilbert, "GM Crop Escapes into the American Wild," *Nature News* 393 (August 6, 2010).

21. C. Greene, S. J. Wechsler, A. Adalja, and J. Hanson, "Economic Issues in the Coexistence of Organic, Genetically Engineered (GE) and Non-GE Crops," Economic Information Bulletin 149, Economic Research Service, U.S. Department of Agriculture, February 2016, 26, https://www.ers.usda.gov/webdocs/publications/44041/56750_eib -149.pdf?v=42424.

22. Greene, Wechsler, Adalja, and Hanson, "Economic Issues in the Coexistence of Organic, Genetically Engineered (GE) and Non-GE Crops," 28.

23. Marie-Monique Robin, *The World according to Monsanto: Pollution, Corruption, and the Control of Our Food Supply* (New York: New Press, 2010).

24. European Natural Soy and Plant Based Foods Manufacturers Association (ENSA), "ENSA Position Paper on GMO and GMO-Free Labeling," February 2013, https://ec.europa.eu/agriculture/sites/agriculture/files/consultations/organic/contri butions/35-ensa_en.pdf.

25. Commission of the European Communities, "Commission Recommendation of 23 July 2003 on Guidelines for the Development of National Strategies and Best Practices to Ensure the Co-existence of Genetically Modified Crops with Conventional and Organic Farming," July 23, 2003, 2.

26. Alexandra Hozzank, "Sustainable Agricultural Systems and GMOs: Is Co-existence Possible?," in *Biological Resource Management in Agriculture: Challenges and Risks of Genetically Engineered Organisms*, ed. Organisation for Economic Co-operation and Development, 161–170 (Paris: OECD Publications, 2004), https://lirias.kuleuven.be /bitstream/123456789/96443/1/OECD+GMOs+Geert+van+Calster+Denise+Prevost .pdf.

27. Clemens C. M. Van De Wiel and L. A. P. Lotz, "Outcrossing and Coexistence of Genetically Modified with (Genetically) Unmodified Crops: A Case Study of the Situation in the Netherlands," *NJAS-Wageningen Journal of Life Sciences* 54, no. 1 (2006): 17–35.

28. Alessandro Chiarabolli, "Coexistence between Conventional, Organic and GM Crops Production: The Portuguese System," *GM Crops* 2–3 (2011): 138–143.

29. Marion Nestle, *Safe Food: Bacteria, Biotechnology, and Bioterrorism* (Berkeley: University of California Press, 2003), 248.

Index

Acquaah, George, *Principles of Plant Genetics and Breeding*, 71
Advanced Genetic Sciences, 36–37
Advanced Genetic Systems (AGS), 61
Agricultural Marketing Act of 1946, 97
Agrobacterium tumefaciens (*A. tumefaciens*), 12–14, 32, 51, 69, 72, 74, 82, 119
  T-DNA in, 12
  Ti plasmid in, 13
  VIR genes in, 12
Akay, Mehmet Turan, and Bt maize study, 111–112
Alliance for Bio-Integrity et al. v. Shalala, 94
Altieri, Miguel, A., on coexistence of GMOs and non-GMOs, 150
American Journal of Clinical Nutrition, study on Golden Rice, 122–123
American Meat Institute, on labeling, 98
Animal feeding experiments, xix, 76, 104, 115, 142–143
Antisense technology, 30, 32–34
Antoniou, Michael N., 65, 81, 84, 88
Asilomar Conference (1975), 26
Atlihanm, Neslihan, 25
Aventis Crop Science, and StarLink corn, 61–63

Bache, Alexander Dallas (first president of National Academy of Sciences), 103
Bacillus thuringiensis (Bt), 58–66, 106, 145
  Cry proteins of, 59, 146
  δ-endotoxins in, 60, 147
  discovery of, 58
  against *Lepdoptera*, 60
  as microbial pest control agent, 60
Bacon, Francis, *The New Atlantis*, xvi
Barrows, Geoffrey, 146
Bartholomaeus, Andrew, 76
Benbrook, Charles M., and pesticide use in U.S., 44, 147
Bennett, Alan B., 146
Berliner, Ernst, 58
Beta carotene
  biofortified rice with, 127
  converted to vitamin A, 126
  in Golden Rice, 119, 121, 126
Beyer, Peter (University of Freiberg, Germany), 120, 123
  in *Science* (magazine), 122, 124
Bingham, John, 1
Binimelis, Rosa, and STS approach to GMOs, 130
Biohegemeny, 135–136
Biolistics, 13, 82. *See also* Gene gun
Biotechnology, agricultural, xv
  Domestic Policy Council Working Group on, xvii

Biotechnology (cont.)
early developments in, xv
oversight of, xvii
Bittman, Mark, on labeling, 100
Bohr, Niels, and model of the atom,
  67
Bondy, Genevieve, 76
Bradford, Kent, J., 70
Brévault, Thierry, on Bt resistance,
  147
Brown, Nancy Marie, 22, 33, 73,
  83
Mendel in the Kitchen, 25, 71, 82
Brumer, Eric, 84
Burbank, Luther, 82
Burkhardt, Peter, 121

Cai, Zongwei, 54
Calgene, 29, 30, 32–34
Camerer, Rudolf Jakob, 3
Carrière, Yves, on Bt resistance, 147
Carson, Rachel, *Silent Spring*, 39
Casman, Elizabeth A., 40
Cauliflower mosaic virus (CaMV35S), as
  promoter sequence, 11, 13, 24, 32.
  See also Promoter
Cayford, Jerry (Resources for the
  Future), 134
Center for Disease Control and
  Prevention (CDC), 64
Central Dogma in genetics, 4
Charney, Evan, 68
Chiarabolli, Alessandro, on Portuguese
  coexistence regulations, 152
Chi-Ham, Cecilia, 146
Chilton, Mary, 13
Cho, Myeong-Je, 73
Choi, Hae-Woon, 73
Ciliberto, Federico, 44, 147
Cisgenesis, 15–16, 25–26. *See also* Crop
  breeding
Clemens, C. M., on coexistence of
  GMOs and non-GMOs, 152

Coat protein gene-mediated transgenic
  resistance, 52
Cockburn, Andrew, 84, 86
Codex Alimentarious Commission,
  United Nations, 74, 114
Colchicine, 7
Coll, Anna, 23
Confirmation bias, xxii
Connecticut, on labeling GMOs, 95
Consumers Union, 22. *See also* Hansen,
  Michael
Council of Europe, on meaning of
  GMO, 81
CRISPR (Clustered regularly interspaced
  short palindromic repeats), 8, 15–16,
  18
CRISPR/Cas9 (gene editing) 18–19,
  24–25, 93, 104, 127
Crop breeding, artificial selection, 2
biofortified crops, xxii, 121, 123, 127,
  149
cell fusion (somatic hybridization),
  7, 25
chromosome engineering, 4
cisgenesis, 15–16, 25–26
cross-breeding, xxi
culturing plant cells, 4, 12
embryo rescue, 4–5
hybridization, 3, 4, 7–8, 12
hybrid seed technology, 3
intragenesis, 15
marker-assisted selection, 8
molecular breeding, xxi, 5, 8–11,
  14, 19–27, 34, 70, 80–82, 141–142,
  148
mutagenesis (radiation or chemical),
  4, 7, 11, 16, 18, 26
outbreeding (outcrossing), 3
pathogens, 50
pure lines, 3
RNA-dependent DNA methylation
  (RdDM), 15, 17
somaclonal variations, 4

synthetic DNA, 15
traditional breeding, xxi, 5, 8, 18–27, 73, 80–82, 141
wide crosses, 5
Cross-protection strategy, for papaya, 51
inducible defenses, 52
pathogen derived resistance, 55
Crown gall disease, 14

Daffodils, as source of beta-carotene, 120, 122
Dairy Food Industry v. Amestoy, on labeling bovine growth hormone, 99
Defarge, Nicolas, 46
Deng, Jianchao, 54
Diamond v. Chakrabarty, 1980, 134
Diels, Johan, 136
DNA Plant Technology, merged with Advanced Genetic Sciences, 37
DOW Agro Sciences, 64
Drake, Pascal M.W., 86
Dutch Coexistence Committee, and commingling of GMO and non-GMO crops, 152

Ecological Society of America, on transgenic canola talk, 151
Electroporation, 7, 13
Enlist Duo herbicide, 44
Environmental Defense Fund, xvi
Environmental Protection Agency (EPA), 42
oversight of biotechnology, xvii, 32
papaya, 52
Scientific Advisory Panel, 63–64
StarLink and, 61
Epigenetics, 68
EPSPS (5-enolpyruvylshikimate-3-phosphate synthase), enzyme that confers tolerance to glyphosate, 41, 85
shikimate pathway, 85
Escherichia coli, kan(r) gene from, 32

Ethylene, treatment of tomatoes with, 31
European Commission, 14
on commingling of GMO and non-GMO crops, 152
European Food and Safety Authority (EFSA), 75, 87, 91–92, 109–111, 117
European Parliament, on labeling GMOs, 94
European Patent Office, 134
European Union
on defining GMOs, 15, 20, 75, 94–95
traceability and labeling regulations in 2003, 97
Excitatory postsynaptic potential synthase (EPSP), enzyme in plants, 85

Fagan, John, 81, 84, 88
Federal Insecticide, Fungicide, and Rodenticide Act (FIFRA), 61
Federoff, Nina, 22, 33, 73, 83
Mendel in the Kitchen, 25, 71, 82
Fernandez-Cornejo, Jorge, 43
Food and Agricultural Organization (FAO), United Nations, 145
crop mutant variety database 7
food safety standards, 114
Food and Chemical Toxicology, retracted paper in, 109
Federal Advisory Committee Act (FACA), conflicts of interest and, 117
Flavr Savr tomato, xxi, 29–35
Food and Drug Administration (FDA), 33, 87
GMO policy, 1992, xvii, 87–88, 94
on labeling, 96
oversight of biotechnology, xvii
StarLink, 61
unexpected outcomes of GMOs, 91
unintended consequences, 141
Food biotechnology, science of, ix
Food Quality Protection Act of 1996, 61

Foundations on Economic Trends, and opposition to GMOs, xvii
Friends of the Earth, and StarLink, 63
Frost Technology Corporation, and Ice Minus, 36
Fuchs, Marc, 53

Gene-drive technology, 37
Gene gun, 11, 51–2, 69. *See also* Biolistics
Generally regarded as safe (GRAS), and GMOs, xvii, 75, 84, 88, 91, 94
Gene stacking, or pyramiding, 44
Genetically modified organisms (GMOs), 139
  activists, xxi
  allergenicity, 87
  animal feeding experiments, xix, 73, 76
  biodiversity of, xx, 146
  biofortification of, xx
  commercial speech of, 98
  debates over, xviii
  deniers of, 79
  environmental impacts of, xx, 20
  Golden Rice, xi, 119–127, 140
  health effects, 72, 79–92
  herbicide resistance, Basta, 134, 145
  introduction of, 9
  labeling of, xx, 93–101
  microbiome, 89
  nutritional quality of, xx, 10, 75
  politics of, ix, xi
  regulation, process-based, xx, 19, 94
  regulation, product-based, xx, 19, 94
  risks of, xix–xxii, 20, 26, 38, 70, 72–74, 98, 130–131, 134, 139
  science of, xi
  scientific consensus, xvii
  systematic reviews of, 116
  yeast, 89
Genetics ID, 63
GGPP (geranylgeranyl-pyrophosphate), precursor to beta carotene, 121

Glycoalkaloids, in potatoes, 107
Glufosinate, 42
  phosphinothricin in, 42
Glyphosate, 39–47
Golden Rice Humanitarian Board, 123
Gongke, Li, 54
Gonsalves, Dennis D. (Cornell University), 51, 53
Green, Jerry M., 45
Greenpeace, International campaign against GMOs, xvi
  Golden Rice, 124
Green Revolution, and Rockefeller Foundation, 120
Grocery Manufacturers Association (GMA), 95–6
Gurian-Sherman, Doug, 42

Halfhill, Matthew D., 53
Hansen, Michael (Consumers Union), 22
Harwood, Wendy, 21
Hefferon, Kathleen L., on gene-edited crops, 127
Hennessy, David, 44
  GMOs and pesticide use, 147
Herrero, Amaranta, and STS approach to GMOs, 130
Herring, Ronald, 131, 133
Heteroencapsidation, 53
Hoechst Schering AgroEvo, 61
Hou, Hongwei, 25
Hub, D., 82
Hudson, Malcolm D., 91

Ice minus, 29, 35–37
  first field test of, 36
  Frostban, 37
  ice nucleation active (INA), 36
  Monterey County, CA, prohibits field tests of, 36
Inducible microbial proteins, 50

Inose, Tomoko (Research Institute for Food Sciences, Kyoto University), 89
Institute of Food and Agricultural Sciences, University of Florida, toxicology of pesticides, 45
Institute of Science in Society, on Golden Rice, 125
International Agency for Research on Cancer (IARC), on reclassified glyphosate, 46
International Atomic Energy Agency, and crop mutant variety database, 8
International Journal of Molecular Sciences, on crop improvement, 149
International Rice Research Institute (IRRI), 125–126
International Service for the Acquisition of Agri-biotech Applications, on biofortification, 127
Intragenesis, 15–16. See Crop breeding

Jasanoff, Sheila, *Design on Nature: Science and Democracy in Europe and the United States*, 130
Jean Mayer USDA Human Nutrition Research Center on Aging, Tufts University, clinical trial on Golden Rice, 122
Jiao, Zhe, 54
Johnson, Katy L. 87, 91
J. R. Simplot Company, and GMO potato, 80

Kanamycin (antibiotic), 32–33
resistance to, 53
Keller, Evelyn Fox, *The Century of the Gene*, 68
Key, Suzie, 86
Keyworth, George, as presidential science adviser, xvii
Kiliç, Aysun, and Bt maize study, 111–112
Kingsbury, Noel, *The History and Science of Plant Breeding*, 7

Kleter, G. A., 82
Ko, E. J., 82
Kramer, Matthew G., 33
Kuhn, Thomas, *The Structure of Scientific Revolutions*, 129
Kurper, H. A., 82

Labeling, 93–101
ballot initiatives on, 99
commercial speech of, 98–99
food package symbols on, 101
laws for GMOs, 95
point of origin, 98
U.S. policy on, 98
Landrace, 2
Latham, Jonathan R., 14, 73
Le Corbusier, *Radiant City*, xvi
Lectins, as insecticidal plant proteins, 58
Lemaux, Peggy, 73
Lindow, Steven E., 35–36
Liu, Biao, 60
Livingston, Mike, 43
Lotz, A. P., on coexistence of GMOs and non-GMOs, 152
Lu, Zhen-Xiang, 25
Lurquin, Paul, *High Tech Harvest: Understanding Genetically Modified Food Plants*, 85

Ma, Julian K.C., 86
MacGregor, and Flavr Savr tomato brand, 33
Maine, on GMO labeling, 95
Marker gene, 11, 13, 23, 73, 80
antibiotic resistant, 11, 52, 98
herbicide resistant, 11
NPTII (neomycin phosphotransferase II), 55
Martineau, Belinda, *First Fruit: The Creation of the Flavr Savr Tomato and the Birth of Biotech Food*, 32, 34
May, Robert, 73
Mayer, Sue, 84

McMichael, Philip, 132
McClintock, Barbara, 69
Mereno-Fierros, Leticia, 65
Merton, Robert, on organized
    skepticism, 141
Mesnage, Robin, 144
Messeguer, Joaquima, 23
Milkweeds
  disappearance of, 46
  Monarch butterflies, 46
Millstone, Erik, 84
Mitchell, Lorraine, 43
Molecular breeding, xi, xix, xxi–xxii, 5, 8,
    9–18, 19–27, 34, 67, 70–73, 76, 79–84,
    90, 98, 105–106, 119, 121–130, 132,
    134, 139, 141–142, 148, 150
Monsanto, 75, 131, 152
  Bt potatoes, 60
  Cockburn at, 84
  discovery of EPSPS enzymes, 41
  glyphosate patented, 40
  herbicide resistance technology, 41–42
  new metabolic pathway, 86
  safety of Roundup, 85
  supporters of, 109
  technology stewardship agreement of,
    132–133
  toxicology of Bt crops, 64
Montero, Maria, 23
Morse, Stephen P., 21
Moschini, GianCarlo, 44
  GMOs and pesticide use, 147
Motta, Renata, social disputes over
    GMOs, 136
Mumm, Rita, H., 21
Murata, Kousaku (Research Institute for
    Food Sciences, Kyoto University), 89

Nadal, Anna, 23
National Academies of Sciences,
    Engineering and Medicine (NASEM),
    21, 72, 142–143
  Academy Forum, 1977, 20

formerly National Academy of
    Sciences (NAS), xvii, 44
  integrity of, 117–118
  National Academy of Engineering,
    103
  National Institute of Medicine, 103
  National Research Council, 103, 105,
    116
  Proceedings, 103
  report on GE crops, xviii, 19, 71, 90,
    98, 103–118
  reports on biotechnology, 104
*National Bioengineered Food Disclosure
    Standard* (S.764), 96
National Research Council (NRC),
    2, 5
Nestle, Marion (New York University),
    Safe Food: The Politics of Food Safety,
    ix–x, 120, 153
Newell, Peter, on biotechnology in
    Argentina, 135
New York Times, op-ed, 93
Nielsen, Kaare, M., 17
Noteborn, J. M., 82
No-till agriculture, 40

Obama, Barack, and National
    Bioengineered Disclosure Standard,
    PL 114-216, 100
ODM (oligonucleotide-directed
    mutagenesis), 15–16. See Crop
    breeding
Office of Technology Assessment (OTA),
    on Ice Minus, 37
OMICS
  compositional analysis of GMOs, 76,
    90, 113–114, 143–144
  genomics, 114, 143
  metabolomics, 77, 90, 114
  proteomics, 65, 90, 114, 143
  transcriptomics, 114, 143
Organic Food Production Act of 1990,
    100

Organisation for Economic Co-operation
and Development (OECD)
*Agricultural Policies in OECD Countries:
Monitoring and Education 2000*, 83
GMO testing guidelines, 109

Paarlberg, Robert, 131, 133
Papaya ringspot virus (PRSV), 50–51, 54
disease resistant, 52
on Oahu Island, Hawaii, 50
Papaya trees, 50
solid properties of, 54
transgenic disease resistance of, 54
Parrott, Wayne, 76
Patenting, of living organisms, 133–134
Pectins, 30
Pellegrini, Pablo, GMOs in Brazil, 136
Perry, Edward D., on GMOs and
pesticide use, 44, 147
Pesticides, early use of, 57–58
Phytoalexins, plant protective proteins,
49
Pia, Maria, 23
Plant Genetic Systems (PGS), 61–63, 134
Plant genome
ecosystem model of, xxii, 68–70
Lego model of, xxii, 67–68
Pleiotropic effects, 24, 82
Plough, Alonzo, *Environmental Hazards*,
27
POEA (polyoxyethyleneamine), as
adjuvant in the herbicide Roundup,
46
Pollen tube pathway (PTP), 14
Polyethylene glycol (PEG), 14
Polygalacturonase (PG), 30–31
pectin degrading, 31
Polymerase chain reaction (PCR), 10
Polyploidy, 8
Popper, Karl, on falsification, 89–90
Poppy, Guy M., 87, 91
Position effect, 68
Post translational modification, 24

Potrykus, Ingo (Swiss Federal Institute
of Technology [SFIT], Zurich),
119–127
in *Science* (magazine), 122
Promoter (gene sequence), 10–11, 13,
23–24, 80
cauliflower mosaic virus (CaMV35S),
11
Erwinia uredovora (soil bacterium), 122
Pseudomonas syringae, 35
Puchta, Holga, 18

Ramazzini Institute, Italy, glyphosate
and the microbiome, 86
Raybould, Alan F., 87, 91
Recombinant DNA Molecule Advisory
Committee (RAC), NIH
review of ice minus, 36
Recombinant DNA technology, 68
pathogen derived resistance, 51
for plant breeding, 9, 22–23
reports on use of, xvii
Redenbaugh, Keith, 33
Restriction enzymes, 10
RNA interference also RNA silencing,
17, 52, 55
Roberts, Richard (New England Biolabs),
124
Robinson, Claire, 65, 81, 88
Roundup, 41, 87
Roush, Richard T., 60
Rubio-Infante, Nestor, 65

Safe and Accurate Food Labeling Act of
2015, (H.R. 1599), 96
Salquist, Roger (Calgene), 33
Schubert, David (Salk Institute), 69,
71
Science
biases in, xxi
conflict of interest in, xxi
Science and Technology Studies (STS),
129–137

Science (magazine), and Golden Rice, 125
Séralini, Gilles-Eric, 109–110
Sexten, Steven, 46
Shao, Jian-Zhou, 60
Shelton, Anthony M., 61
Shigetane, Ishiwata, and isolated
    Bacillus thuringiensis, 58
Shiva, Vandana, *Biopiracy: The Plunder of
    Nature and Knowledge*, 133
Silicon carbide mediated transformation
    (SCMT), 14
Small, Mitchell J., 40
Snow, C. P., ix, x
StarLink, genetically modified corn,
    61–63
Steinbrecher, Ricarda A., 73, 82
Stephan, Hannes, 27
Stewart, C. Neal, 53
Substantial equivalence, 75, 83–85, 89,
    90, 107, 139, 144
Sungenta, 124
Sustainable agriculture, xx, 148–149
Systematic acquired resistance (SAR), 49

Tabashnik, Bruce E., on Bt resistance, 147
Tan, Yanhua, 144
Tang, Guangwen, 124
Taxonomy of living organisms, 6
Termination sequence (stop codon), 11,
    13, 23, 80
Terpenoids, or terpenes, plant chemicals
    responsible for smell and color, 120
Tobacco mosaic virus (TMV), 52
Toenniessen, Gary (Rockefeller
    Foundation), 120
Toxicity studies, and whole food on
    animals, 76
Traditional breeding. *See* Crop breeding
Transgene complex or cassette, 9–14,
    23, 32, 51, 74, 122
  carried by vectors, 12–13
  cells, 10
  introgression, 54

Transgenic crops, or genetically
    engineered crops and genetically
    modified crops, xv, 14, 17
  famigenic, 17
  first generation, xvi
  imprecise, 24
  intragenic, 17
  linegenic, 17
  unpredictable, 24
Triazine, family of herbicides, 40
2,4-D, herbicide, cited in *Science*
    (magazine), 39

Union of Concerned Scientists, xvi, 42
University of California, Berkeley, and
    ice nucleation, 35
University Genetics Company
    (Norwalk, CT), 56
University of Wyoming, and ice
    nucleation, 35
U.S. Department of Agriculture (USDA),
    42
  animal feed data, 112
  Animal Plant Health Inspection
    Service (APHIS), 52
  approved GMO potato, 80
  Coordinated Framework for
    Regulation of Biotechnology (1986),
    32
  Economic Research Service (ERS), 151
  on glyphosate resistance, 40
  on herbicide use 43
  on labeling, 96, 101
  national organic survey, 15
  Regulation of Biotechnology (1986), 32.
    StarLink, 61
  on yields of herbicide-resistant crops,
    42

Vaeck, Mark, 61
Valentine, Ray, 30
Van Asselt, Marjolein B. A., on
    uncertain risks of GMOs, 131

Van De Wiel, Clemens C. M., on
coexistence of GMOs and non-
GMOs, 152
Van Eenennaam, Alison L., farm animal
study by, 112
Vaucheret, Hervé, 69
Vectors, for transporting genes
bacteria (*Agrobacterium tumefaciens*), 12
plasmids, 112
viruses, 12
Vermont
on labeling bovine growth hormone,
99
on labeling GMOs, 95, 98
Vitamin A deficiency, or
hypovitaminosis, 125
biofortified rice, 121
Golden Rice, 119
world health problem, 119
Volunteer plants, 40
Von Kraus, Martin Paul Krayer, 40
Vos, Ellen, uncertain of risks of GMOs,
131

Walker, Kate, 76
Walter, Felix, 18
Warwick, Suzanne I., 53
Wechsler, Seth, 43
Weeds, herbicide resistant, x, 44
Weight of evidence, 76, 139
Wei, Xiang-dong, 54
Wessler, Justus, on economics of Golden
Rice, 125
Wickso, Fern, on STS approach to
GMOs, 130
Wilson, Allison K., 73, 82

Xenogenic, 17
Xue, Kun, 60, 65

Yang, Jing, 60
Ye, Xudong, and beta carotene rice,
121

Yearley, Steven, on STS approach to
GMOs, 129
Young, Andrea E., farm animal study
by, 112

Zakharyan, Robert, on isolated plasmids
in *Bacillus thuringiensis*, 59
ZFNs (zinc finger nucleases for gene
editing), 15
Zhang, Zhuomin, 54
Zilberman, David, 146
on economics of Golden Rice, 125
Zue, Dayuan, 60

Other books authored, coauthored, or coedited by Sheldon Krimsky

Genetic Alchemy: The Social History of the Recombinant DNA Controversy (1982)
Environmental Hazards: Communicating Risks as a Social Process (1988)
Biotechnics and Society: The Rise of Industrial Genetics (1991)
Social Theories of Risk (1992)
Agricultural Biotechnology and the Environment (1996)
Hormonal Chaos: The Scientific and Social Origins of the Environmental Endocrine Hypothesis (2000)
Science and the Private Interest (2003)
Rights and Liberties in the Biotech Age (2005)
Genetic Justice: DNA Databanks, Criminal Justice and Civil Liberties (2011)
Race and the Genetic Revolution: Science, Myth and Culture (2011)
Genetic Explanations: Sense or Nonsense (2013)
Biotechnology in Our Lives (2013)
The GMO Deception (2014)
Stem Cell Dialogues: A Philosophical and Scientific Inquiry into Medical Frontiers (2015)
Conflict of Interest in Science (2018)